T0360413

Optimal Fractionation in Radiotherapy

This monograph provides a mathematically rigorous overview of optimal fractionation in cancer radiotherapy, which seeks to address the fundamental trade-off of maximizing tumor-kill while protecting nearby healthy tissue from toxic effects. Most mathematical research on exact solutions to this problem is scattered across journals in applied mathematics, operations research, radiobiology, medicine, and medical physics. These works rarely include rigorous proofs or detailed derivations of their methodologies. Moreover, computer programs utilized for conducting numerical experiments seldom accompany these publications, thereby jeopardizing reproducibility. This monograph aims to provide a comprehensive pedagogical reference that brings researchers up to speed on optimal fractionation, utilizing and illustrating analytical techniques from linear algebra, calculus, linear programming, quadratic and nonlinear programming, robust optimization, and inverse optimization. Its purpose is to help readers understand the mathematics behind the optimal fractionation problem, empowering them to contribute original work to this field.

ARCHIS GHATE is a professor of Industrial & Systems Engineering at the University of Washington, where he previously served for five years as the College of Engineering Endowed Professor of Healthcare Operations Research. His research focuses on optimization. He received a PhD in Industrial and Operations Engineering from the University of Michigan in 2006 and an MS in Management Science and Engineering from Stanford University in 2003.

"This book provides a wonderfully rigorous and comprehensive treatment of the mathematics behind fractionation in radiation therapy. It consolidates a significant body of work in a consistent manner, making it a go-to reference for the expert and beginner alike. The inclusion of robust optimization and inverse optimization in the context of fractionation provides considerable depth, reinforcing the value of interdisciplinary research."

Timothy Chan, *University of Toronto*

"This text provides a unifying mathematical framework for the fractionation decision in radiotherapy. It offers a rich blend of mathematical analysis, numerical examples, codes, exercises, and bibliographic notes for an in-depth understanding and implementation of the concepts discussed. The book presents a fresh perspective to the long-standing fractionation debate in the field of radiotherapy and can inspire the design of new clinical trials to improve treatment efficacy for cancer patients."

Ehsan Salari, *Wichita State University*

Optimal Fractionation in Radiotherapy

ARCHIS GHATE
University of Washington

Shaftesbury Road, Cambridge CB2 8EA, United Kingdom

One Liberty Plaza, 20th Floor, New York, NY 10006, USA

477 Williamstown Road, Port Melbourne, VIC 3207, Australia

314–321, 3rd Floor, Plot 3, Splendor Forum, Jasola District Centre,
New Delhi – 110025, India

103 Penang Road, #05–06/07, Visioncrest Commercial, Singapore 238467

Cambridge University Press is part of Cambridge University Press & Assessment,
a department of the University of Cambridge.

We share the University's mission to contribute to society through the pursuit of
education, learning and research at the highest international levels of excellence.

www.cambridge.org
Information on this title: www.cambridge.org/9781009341134

DOI: 10.1017/9781009341110

First published 2024

A catalogue record for this publication is available from the British Library

Library of Congress Cataloging-in-Publication data
Names: Ghate, Archis, author.
Title: Optimal fractionation in radiotherapy / Archis Ghate.
Description: New York : Cambridge University Press, 2023.
Identifiers: LCCN 2023011820 (print) | LCCN 2023011821 (ebook) | ISBN 9781009341134
(hardback) | ISBN 9781009341110 (epub)
Subjects: MESH: Dose Fractionation, Radiation | Dose-Response Relationship, Radiation
Classification: LCC RM849 (print) | LCC RM849 (ebook) | NLM WN 250.6 |
DDC 615.8/42–dc23/eng/20230530
LC record available at https://lccn.loc.gov/2023011820
LC ebook record available at https://lccn.loc.gov/2023011821

ISBN 978-1-009-34113-4 Hardback

Additional resources for this publication at www.cambridge.org/9781009341134/Resources

To my PhD students.

Contents

Preface

This monograph is a detailed expository adaptation of some of the papers previously published by my research group on the optimal fractionation problem in cancer radiotherapy. Although this problem has been known to radiation treatment planners for more than a century, its mathematical formulations and solutions have been derived more recently. These are predominantly rooted in the so-called linear quadratic dose-response model from radiobiology. The monograph provides detailed mathematical derivations of these formulations and corresponding solution methods. The focus is mostly on exact analytical solutions; that is, solutions that can be expressed as closed-form formulas. Special attention is paid to structural properties of these solutions and their interpretations. In a few cases where a closed-form formula is not available, a solution based on a linear program or a quadratically constrained linear program is computed. A pseudocode listing is provided for each solution method. These were largely missing in the original publications, and I hope that they will help readers better understand the mathematical methods. Accompanying Python computer programs and data are also provided as supplementary materials available at www.cambridge.org/9781009341134/Resources. The code was written with easy readability in mind rather than runtime efficiency. Qualitative insights are drawn from numerical experiments throughout. The readers can attempt to reproduce these experiments using the given Python code, conduct further experiments to enhance their understanding, or modify the code as appropriate. Since I wrote and debugged all code myself, it must not be used in any real clinical applications as it has not undergone sufficient external scrutiny.

A few exercises are included at the end of each chapter. Some of the exercises request the readers to perform additional numerical experiments. Some exercises are open-ended (marked with a star), while others request readers to develop variations of the solution methods discussed in the chapters. I do not know how to solve the open-ended questions. So I call upon readers to inform me via email if they find answers to those questions.

Readers are assumed to be familiar with calculus, linear algebra, and basic concepts from linear and nonlinear optimization at the undergraduate level. An appendix on background in optimization is included. I recommend reading the chapters in the order presented, as each chapter builds upon the previous one. My hope is that this monograph will serve as a compendium for readers who are new to this topic as well as for those who are somewhat familiar with it. It could also be used as a textbook for a 10-week course, covering one chapter in one week on average.

I am grateful to several former PhD students who coauthored papers with me that laid the foundation for this monograph. Some of the figures in this monograph are adapted or reproduced from our previously published work. A note of appreciation is due to the publishers for allowing me to do this. I am thankful to the University of Washington, Seattle, for supporting my sabbatical leave in the academic year 2021–2022. This monograph was written during that sabbatical.

Finally, I encourage readers to convey to me via email any typographical errors, and other mistakes in the text as well as in the Python code.

1
Introduction

The goal in external beam radiotherapy (see Figure 1.1) for cancer is to kill the tumor while limiting toxic effects of radiation on nearby organ(s)-at-risk (OAR). At least three avenues can be pursued to attain this goal: spatial localization of radiation dose (energy per unit mass); temporal dispersion of radiation dose; and selection of an appropriate radiation modality to administer the dose.

In spatial localization, a high dose is prescribed to the tumor, while upper limits are recommended on doses to the OAR. The fluence-map optimization problem then seeks a radiation intensity profile that meets this protocol as closely as possible. Modern technology such as intensity-modulated radiation therapy (IMRT) can be employed to deliver the resulting optimal intensity profile. See Figure 1.2.

Healthy cells are better at recovering from radiation damage than tumor cells. Thus, the planned total dose is administered over multiple treatment sessions. This is called fractionation, and it gives healthy cells some time to recover between sessions, thereby reducing the overall toxic effect. This might suggest that the larger the number of sessions, the better. However, the tumor can grow over the treatment course, and thus it is crucial to eliminate viable tumor cells with a short course. Moreover, in some cases, a short course could be more effective, even without tumor proliferation, because the tumor's response to radiation is similar to that of a nearby OAR. Treatment planners are thus interested in finding an optimal number of treatment sessions and also the dose in each session. This is called the optimal fractionation problem. The differences between the radiobiological response of the tumor and various OAR are at the heart of this temporal problem. See Figure 1.3.

Radiotherapy can be administered via different modalities such as photons and protons. The choice of a modality may depend on the cancer anatomy;

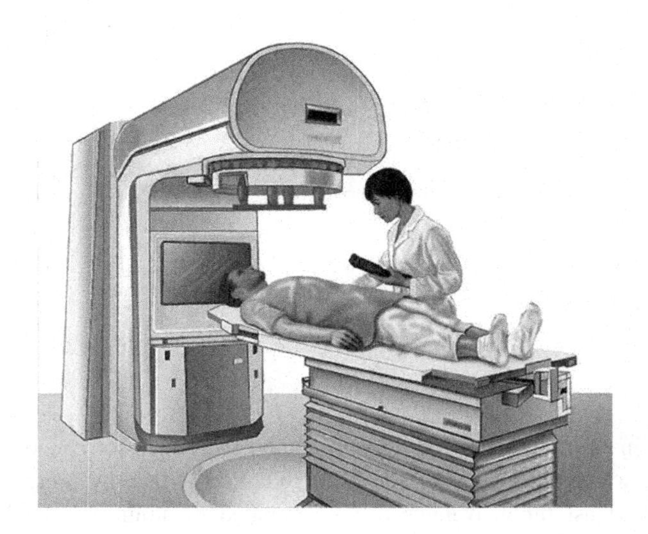

Figure 1.1 A linear accelerator machine for external beam radiotherapy. Radiation beams emerge from the top of the machine and pass through the patient's body. Source: National Cancer Institute, www.cancer.gov/about-cancer/treatment/types/radiation-therapy/external-beam and https://visualsonline.cancer.gov/details.cfm?imageid=9413. Reused per policy available at www.cancer.gov/policies/copyright-reuse.

physical properties, such as the so-called dose-deposition profile, of the modalities (see Figure 1.4); biological properties, such as the relative dose sensitivities of the tumor and OAR to these modalities; and the capital investment or operating cost of administering treatment with these modalities. A modality that is superior from one perspective under certain disease conditions may be inferior under different disease conditions or from a different perspective. As such, there is no universally dominant modality. These trade-offs between modalities are further complicated when a treatment planner attempts to determine the number of treatment sessions and the doses that should be administered via each available modality. This can be viewed as optimal fractionation with multiple modalities.

This monograph describes mathematical optimization models and solution methods for the fractionation problem with one and two modalities. All optimization models in this monograph are based on the linear-quadratic (LQ) framework of dose response. According to this framework, the damage caused by radiation to the tumor or an OAR is modeled using two components. The first component is linear in dose whereas the second is quadratic in dose. Tumor proliferation is modeled using a separate, third component that

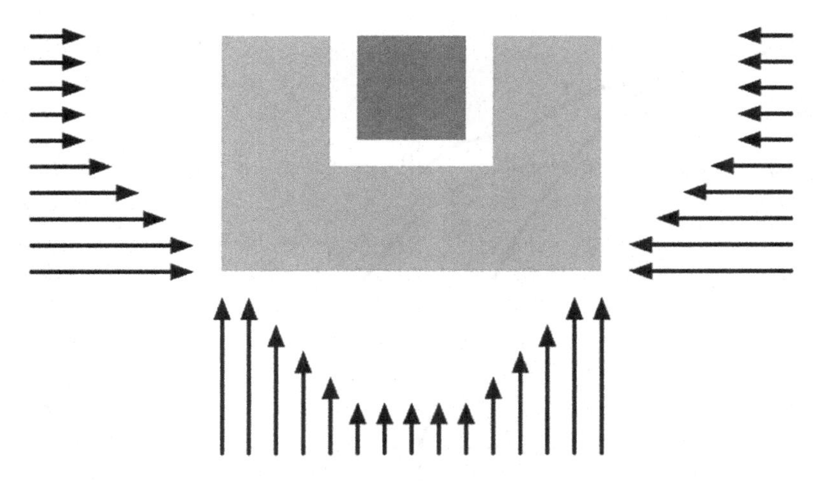

Figure 1.2 The fluence-map problem seeks a radiation intensity profile (black arrows) that administers a high dose to the tumor (light gray U shape) and a low dose to the OAR (dark gray square). The lengths of the arrows represent radiation intensity. Radiation beams from multiple directions (three here) are employed to ensure that all parts of the tumor receive sufficient dose.

depends on the length of the treatment course. According to this framework, the surviving fraction of tumor cells is modeled as

$$\exp\left(-\alpha_0 \sum_{t=1}^{N} d_t - \beta_0 \sum_{t=1}^{N} (d_t)^2 + \tau(N)\right). \tag{1.1}$$

Here, d_t is the dose administered to the tumor in session t; N is the number of sessions; $\alpha_0 > 0$ and $\beta_0 > 0$ are parameters; and $\tau(N)$ is the proliferation term that depends on the number of sessions N. This LQ dose-response model is simple and has been validated with data. It is commonly employed for calculating and comparing effects of competing dosing plans. This monograph relies on the simplicity of this framework to derive insights into solutions of various optimization models via an interplay between algebra, geometry, and calculus.

Chapters 2–5 assume that a radiation intensity profile is determined a priori; the decision-maker only needs to decide how to disperse the resulting total dose across treatment sessions. This can be viewed as spatiotemporally separated fractionation, and the corresponding optimization problems can be solved exactly. This could be suboptimal as compared to simultaneously determining both the intensity profile and the dose dispersion plan. This

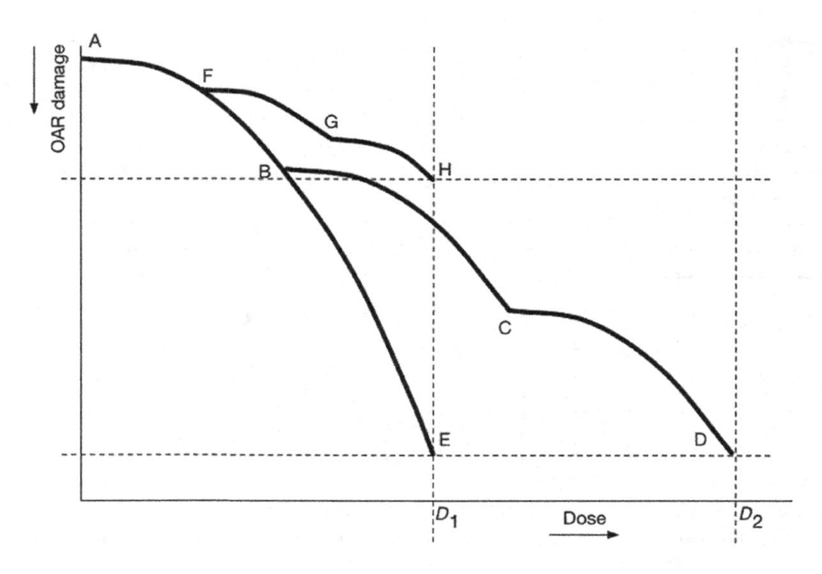

Figure 1.3 The damage to the OAR depends on how the total radiation dose is dispersed across multiple treatment sessions. A total dose of D_1 in a single session (curve AE) causes the same damage as a larger total dose of D_2 if it is broken into three sessions (curves AB–BC–CD). Similarly, if the total dose of D_1 is broken into three sessions (curves AF–FG–GH), the damage to the OAR is less than administering D_1 in a single session. Similar curves can also be sketched for the tumor, and the differences between the OAR curves and the tumor curves introduce trade-offs into the optimal fractionation problem. Adapted with permission from [59, Figure 3].

latter, computationally more demanding, approach is termed spatiotemporally integrated fractionation. Chapters 6 and 7 describe mathematical models and approximate solution methods to compute the number of sessions and the intensity profiles in each session for spatiotemporally integrated fractionation. Chapters 8 and 9 describe spatiotemporally separated mathematical models and exact solution methods for optimal fractionation with two modalities. The monograph concludes by outlining directions for future research in Chapter 10.

A comment on terminology: we use increasing to mean nondecreasing and use decreasing to mean nonincreasing throughout. The terms strictly increasing and strictly decreasing are used rarely, only when absolutely necessary. Notation such as $t = 1: N$ is short for the more familiar form $t = 1, 2, \ldots, N$.

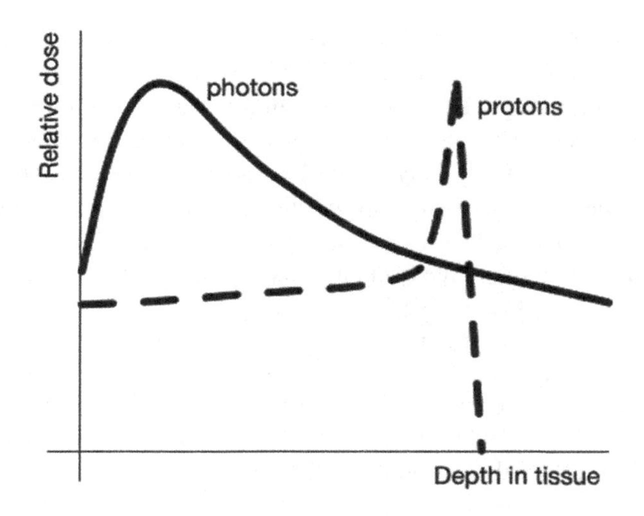

Figure 1.4 A rough schematic of the relative radiation dose deposited as a function of the distance traveled inside a tissue. This is called a dose-deposition profile. The sharp spike in the proton profile is called the Bragg peak. Reused with permission from [61, figure 4].

Bibliographic Notes

Information about IMRT is available in [89, 128]. IMRT is considered by some to be one of the most successful developments in radiation oncology [33]. Others have expressed concerns about its merits. A discussion about the benefits offered and challenges posed by IMRT is included in section 1A of [128], in a point-counterpoint format. Sophisticated models and solution algorithms for fluence-map optimization have been developed over the last three decades [9, 28, 34, 41, 42, 62, 98, 99, 110, 118, 128, 136]. The optimal fractionation problem has a hundred-year history in the clinical literature [3, 10, 16, 20, 54, 58, 68, 69, 70, 85, 97, 100, 120, 134]. Mathematical models of this problem almost exclusively utilize the LQ framework of dose response. A textbook description of the LQ framework and illustrations similar to Figure 1.3 are available, for example, in [65]. The LQ framework was proposed at least as early as the 1940s [83] and has been reviewed repeatedly over the last several decades [24, 35, 46, 52, 73, 94, 106, 111, 119, 124, 130]. Several favorable properties of the LQ framework are listed in [24], stating that "it is reasonably well validated, experimentally and theoretically ...." Estimated values of the parameters in the LQ framework have also been reported [57, 121, 124, 132]. Nevertheless, questions have been raised about its appropriateness for clinical decision-making [26, 80, 137]. An illuminating account of the history; clinical applicability; usage; mechanistic, empirical, and mathematical underpinnings; and concerns about the validity of the LQ framework is available in [86]. This monograph does not take any position on the appropriateness of the LQ framework. It simply provides various mathematical formulations and corresponding solutions of the fractionation problem to guide decision-makers who may wish to utilize the LQ framework. Discussions of the pros and cons of various competing modalities for external beam radiotherapy are included in [38, 40, 64, 66, 67, 125]. Scientifically accurate versions, drawn to scale based on real data/analytical calculations/simulations, of the rough schematic in Figure 1.4 are available, for example, in [27, 66, 81]. These publications also include a technical description of the Bragg peak.

2
Fractionation with a Single Organ-at-Risk

In this chapter, we will study in detail a mathematical model of the optimal fractionation problem with a single OAR, within the LQ dose-response framework. This is typically formulated as

$$(\text{P})\, f_1^\star = \max_{\vec{d},\, N} \; \alpha_0 \sum_{t=1}^{N} d_t + \beta_0 \sum_{t=1}^{N} (d_t)^2 - \tau(N) \tag{2.1}$$

$$s \sum_{t=1}^{N} d_t + s^2 \rho \sum_{t=1}^{N} (d_t)^2 \leq \underbrace{T\delta(1+\rho\delta)}_{C} \tag{2.2}$$

$$\vec{d} \geq 0 \tag{2.3}$$

$$1 \leq N \leq N_{\max}, \text{ integer.} \tag{2.4}$$

The subscript "1" in the optimal objective function value $f_1^\star$ signifies that the problem includes only one OAR. In this problem, $N \geq 1$ is an integer number of treatment sessions, assumed for simplicity to be administered one per day. This is a decision variable. The number of sessions is no more than a given upper bound $N_{\max}$. The real number d_t denotes the dose administered to the tumor in session t. The notation $\vec{d} = (d_1, \ldots, d_N)$ represents a vector of these doses. This is also a decision variable; each component of this vector must be nonnegative. The linear and quadratic response parameters for the tumor are denoted by $\alpha_0 > 0$ and $\beta_0 > 0$, respectively. The term $\tau(N)$ accounts for tumor proliferation over the treatment course. For concreteness, we will utilize the specific form $\tau(N) = \frac{(N-1)\ln(2)}{T_{\text{double}}}$ throughout. Here, T_{double} is the amount of time (days) in which the number of tumor cells doubles. This is called doubling time. The objective function in (2.1) is called the biological effect of the dosing plan $\vec{d} \geq 0$ on the tumor. This biological effect equals the negative of the natural logarithm of the surviving fraction of tumor cells as defined by the LQ framework in (1.1). Thus, the higher the biological effect, the lower the surviving fraction of tumor cells. The goal therefore is to maximize the biological effect. The parameter $s > 0$ on the left-hand side of (2.2) is called the sparing factor. Specifically, sd_t is the dose administered to the OAR in session t. This models the typical situation that if a certain dose is administered to the tumor, then some proportion of that dose is inevitably administered to the OAR. The parameter $\rho > 0$ equals the ratio β/α of the quadratic and linear response parameters $\beta > 0$ and $\alpha > 0$ for the OAR. The expression on the left-hand side of (2.2) is called the biologically effective dose (BED) corresponding to the plan $s\vec{d}$, according to the LQ framework. The right-hand side of that constraint, which we denote by C for brevity, also equals the BED corresponding to a conventional dosing plan that the OAR is known to tolerate. This conventional plan administers a dose of $\delta > 0$ per session in $T \geq 1$ sessions. The constraint ensures that the BED on the left-hand side is no more than the tolerable value on the right-hand side.

We will generalize this formulation and its solution to the more realistic case of multiple OAR in the next chapter. Nevertheless, it is helpful to first focus on the simpler case of a single OAR for ease of exposition. Furthermore, the techniques we will utilize in the next chapter build upon the discussion here. This will allow us to carefully contrast the insights provided by the two chapters.

We will first solve problem (P) for a fixed integer value of $N \geq 1$. We write it as

$$(\mathrm{P}(N)) \quad f_1^\star(N) = \max_{\vec{d}} \; f_1(N,\vec{d}) = \alpha_0 \sum_{t=1}^{N} d_t + \beta_0 \sum_{t=1}^{N} (d_t)^2 - \tau(N) \tag{2.5}$$

$$s \sum_{t=1}^{N} d_t + s^2 \rho \sum_{t=1}^{N} (d_t)^2 \leq C \tag{2.6}$$

$$\vec{d} \geq 0. \tag{2.7}$$

The following proposition completely characterizes all optimal solutions of problem (P(N)), which is a quadratically constrained quadratic program (QCQP).

Proposition 2.1 *We make three separate claims.*

Claim 1 *Suppose* $\frac{\alpha_0}{\beta_0} > \frac{1}{s}\frac{\alpha}{\beta}$. *Then, problem (P($N$)) has a unique optimal solution given by* $d_1 = d_2 = \cdots = d_N = d^\star(N)$, *where*

$$d^\star(N) = \frac{-1 + \sqrt{1 + \frac{4\rho C}{N}}}{2s\rho}. \tag{2.8}$$

We call this an equal-dosage plan.

Claim 2 *Suppose* $\frac{\alpha_0}{\beta_0} < \frac{1}{s}\frac{\alpha}{\beta}$. *Then, the only optimal solutions to problem* $(P(N))$ *are such that the dose in one session equals*

$$d^\circ = \frac{-1+\sqrt{1+4\rho C}}{2s\rho}, \tag{2.9}$$

and the doses in the other $N-1$ *sessions are* 0. *We call these single-dosage plans.*

Claim 3 *Suppose* $\frac{\alpha_0}{\beta_0} = \frac{1}{s}\frac{\alpha}{\beta}$. *Then, any nonnegative dosing plan that satisfies*

$$s\sum_{t=1}^{N} d_t + s^2\rho\sum_{t=1}^{N}(d_t)^2 = C \tag{2.10}$$

is optimal to problem $(P(N))$. *For instance, the equal-dosage plan in* (2.8) *and the single-dosage plan in* (2.9) *are both optimal.*

The condition $\frac{\alpha_0}{\beta_0} > \frac{1}{s}\frac{\alpha}{\beta}$ in the first claim of this proposition is expected to be met for cancers such as head-and-neck and lung. The condition $\frac{\alpha_0}{\beta_0} < \frac{1}{s}\frac{\alpha}{\beta}$ in the second claim is believed to hold for prostate cancer. Indeed, clinical trials with short treatment courses have been conducted for prostate cancer radiotherapy.

This proposition implies that the condition $\frac{\alpha_0}{\beta_0} \geq \frac{1}{s}\frac{\alpha}{\beta}$ is necessary for the optimality of the equal-dosage plan. Similarly, the condition $\frac{\alpha_0}{\beta_0} \leq \frac{1}{s}\frac{\alpha}{\beta}$ is necessary for the optimality of single-dosage plans. We provide five proofs of this proposition as they contain distinct insights.

The problem is easy to solve when $N = 1$. In that case, the problem includes only one variable, which we denote by a real number d. Since the objective function is increasing in this dose variable, constraint (2.6) must be active at an optimal solution. For if not, we could increase d until the constraint becomes active, thereby increasing the objective value. Thus, we have $sd + s^2\rho d^2 = C$. The only nonnegative (in fact, positive) solution of this quadratic equation is $d = \frac{-1+\sqrt{1+4\rho C}}{2s\rho}$, and consequently, this must be optimal. This establishes the proposition for the trivial case of $N = 1$. We therefore need to focus only on the case $N \geq 2$.

2.1 Optimal Solution Using the Cauchy–Schwarz Inequality

As explained earlier, constraint (2.6) must be active at an optimal solution. Thus, we must have (2.10). We use this equation to express $\sum_{t=1}^{N}(d_t)^2 = \frac{C - s\sum_{t=1}^{N} d_t}{s^2\rho}$ and substitute this back into the objective function to obtain

$$\begin{aligned}
&\alpha_0 \sum_{t=1}^{N} d_t + \beta_0 \frac{\left(C - s\sum_{t=1}^{N} d_t\right)}{s^2\rho} - \tau(N) \\
&= \left(\sum_{t=1}^{N} d_t\right)\left(\alpha_0 - \frac{\beta_0}{s\rho}\right) + \frac{\beta_0 C}{s^2\rho} - \tau(N) \\
&= \left(\sum_{t=1}^{N} d_t\right)\beta_0\left(\frac{\alpha_0}{\beta_0} - \frac{\alpha}{s\beta}\right) + \frac{\beta_0 C}{s^2\rho} - \tau(N).
\end{aligned} \tag{2.11}$$

This suggests that the sign of $\left(\frac{\alpha_0}{\beta_0} - \frac{\alpha}{s\beta}\right)$ characterizes optimal solutions. Proposition 2.1 formalizes this idea. Recall the Cauchy–Schwarz inequality, which states that if $u_1, \ldots, u_N$ and $v_1, \ldots, v_N$ are real numbers, then

$$\sum_{t=1}^{N} u_t v_t \leq \sqrt{\sum_{t=1}^{N} u_t^2}\sqrt{\sum_{t=1}^{N} v_t^2}, \tag{2.12}$$

with equality holding only when $u_1 = \cdots = u_N$ and $v_1 = \cdots = v_N$. We now prove the three claims separately.

For Claim 1, if $\frac{\alpha_0}{\beta_0} > \frac{1}{s}\frac{\alpha}{\beta}$, then the objective function as written in (2.11) shows that it is optimal to choose the largest possible value of $\sum_{t=1}^{N} d_t$. Applying the Cauchy–Schwarz inequality (2.12) with $u_1 = \cdots = u_N = 1$ and $v_t = d_t$, for $t = 1\colon N$, demonstrates that the largest possible value of $\sum_{t=1}^{N} d_t$ is $\sqrt{N\sum_{t=1}^{N}(d_t)^2}$ and that this is attained when $d_1 = \cdots = d_N$. We call this value of dose $d^\star(N)$. Substituting this back into (2.10) yields the quadratic equation $Nsd^\star(N) + Ns^2\rho(d^\star(N))^2 = C$. It has a unique positive solution and it is given by formula (2.8).

For Claim 2, if $\frac{\alpha_0}{\beta_0} < \frac{1}{s}\frac{\alpha}{\beta}$, then the objective function as written in (2.11) shows that it is optimal to choose the smallest possible value of $\sum_{t=1}^{N} d_t$. We know that $\left(\sum_{t=1}^{N} d_t\right)^2 \geq \sum_{t=1}^{N}(d_t)^2$ because the doses are nonnegative. That is, $\sum_{t=1}^{N} d_t \geq \sqrt{\sum_{t=1}^{N}(d_t)^2}$. Further, equality holds only when at most one of the doses is positive. Since it is not optimal to set all doses to 0, we call the single positive dose d°. Substituting it back into (2.10) yields the quadratic equation $sd^\circ + s^2\rho(d^\circ)^2 = C$. It has a unique positive solution and it is given by formula (2.9).

For Claim 3, if $\frac{\alpha_0}{\beta_0} = \frac{1}{s}\frac{\alpha}{\beta}$, then the value of the objective function as written in (2.11) does not depend on any doses. In other words, any nonnegative dosing plan that satisfies (2.10) is optimal. Since this third claim is rather trivial, the four other proofs below mostly focus on the first two claims.

2.2 Algebraic Derivation of Optimal Solution

To prove Claim 1, consider any feasible dosing plan $\vec{d} = (d_1, \ldots, d_N)$ wherein there exist at least two distinct sessions j and k such that $d_j \neq d_k$. We call this an unequal-dosage solution. We define the alternative equal-dosage solution $\vec{b} = \underbrace{(b, \ldots, b)}_{N \text{ times}}$, where $b = \sum_{t=1}^{N} d_t/N$. That is, the dose in every session within $\vec{b}$ equals the average of all doses in $\vec{d}$. This dosing plan is strictly feasible because

$$sNb + s^2\rho Nb^2 = s\sum_{t=1}^{N} d_t + s^2\rho N \frac{\left(\sum_{t=1}^{N} d_t\right)^2}{N^2} = s\sum_{t=1}^{N} d_t + s^2\rho \frac{\left(\sum_{t=1}^{N} d_t\right)^2}{N}$$
$$< s\sum_{t=1}^{N} d_t + s^2\rho \sum_{t=1}^{N} (d_t)^2 \leq C.$$

Here, the strict inequality follows by applying the Cauchy–Schwarz inequality (2.12) with $u_t = 1$ and $v_t = d_t$, for $t = 1\colon N$, because not all components of $\vec{d}$ are identical. Strict inequality implies that there exists an $\epsilon > 0$ such that the dosing plan $\vec{w} = \underbrace{(b+\epsilon, \ldots, b+\epsilon)}_{N \text{ times}}$ renders constraint (2.6) active. This means that $sN(b+\epsilon) + s^2\rho N(b+\epsilon)^2 = C$. Consequently, $N(b+\epsilon)^2 = \frac{C - sN(b+\epsilon)}{s^2\rho}$. Thus, the objective value of this plan is

$$\begin{aligned}
f_1(N, \vec{w}) &= \alpha_0 N(b+\epsilon) + \beta_0 N(b+\epsilon)^2 - \tau(N) \\
&= \alpha_0 N(b+\epsilon) + \beta_0 N(b+\epsilon)^2 - \tau(N) \\
&= \alpha_0 N(b+\epsilon) + \beta_0 \left(\frac{C - sN(b+\epsilon)}{s^2\rho}\right) - \tau(N) \\
&= N(b+\epsilon)\left(\alpha_0 - \frac{\beta_0}{s\rho}\right) + \frac{C\beta_0}{s^2\rho} - \tau(N) \qquad (2.13) \\
&= \left(\alpha_0 - \frac{\beta_0}{s\rho}\right) Nb + N\epsilon\left(\alpha_0 - \frac{\beta_0}{s\rho}\right) + \frac{C\beta_0}{s^2\rho} - \tau(N) \\
&= \left(\alpha_0 - \frac{\beta_0}{s\rho}\right)\left(\sum_{t=1}^{N} d_t\right) + N\epsilon\left(\alpha_0 - \frac{\beta_0}{s\rho}\right) + \frac{C\beta_0}{s^2\rho} - \tau(N)
\end{aligned}$$

$$
\begin{aligned}
&\geq \left(\alpha_0 - \frac{\beta_0}{s\rho}\right)\left(\sum_{t=1}^{N} d_t\right) + N\epsilon\left(\alpha_0 - \frac{\beta_0}{s\rho}\right) \\
&\quad + \frac{\left(s\sum_{t=1}^{N} d_t + s^2\rho\sum_{t=1}^{N}(d_t)^2\right)\beta_0}{s^2\rho} - \tau(N) \qquad (2.14) \\
&= \left(\alpha_0 - \frac{\beta_0}{s\rho}\right)\sum_{t=1}^{N} d_t + N\epsilon\left(\alpha_0 - \frac{\beta_0}{s\rho}\right) + \frac{\beta_0}{s\rho}\sum_{t=1}^{N} d_t \\
&\quad + \beta_0\sum_{t=1}^{N}(d_t)^2 - \tau(N) \\
&= \alpha_0\sum_{t=1}^{N} d_t + \beta_0\sum_{t=1}^{N}(d_t)^2 - \tau(N) + N\epsilon\left(\alpha_0 - \frac{\beta_0}{s\rho}\right) \\
&= f_1(N,\vec{d}) + N\epsilon\left(\alpha_0 - \frac{\beta_0}{s\rho}\right) \\
&> f_1(N,\vec{d}).
\end{aligned}
$$

Here, inequality (2.14) follows because $\vec{d}$ is feasible to (2.6). The subsequent strict inequality holds because $\epsilon > 0$ by construction and $\alpha_0 - \frac{\beta_0}{s\rho} > 0$ by assumption. In other words, the alternative equal-dosage plan is feasible and attains a strictly larger objective value than the unequal-dosage plan. This demonstrates that unequal-dosage plans are not optimal. We must, therefore, restrict attention only to equal-dosage solutions. This proves the claim because the best equal-dosage solution is given by formula (2.8).

We prove Claim 2 by contradiction. Suppose dosing plan $\vec{d}$ wherein there exist at least two sessions with positive doses is optimal. Specifically, let $\mathcal{T}$ be the set of sessions with positive doses. That is, $\mathcal{T} = \{t \in \{1\colon N\} | d_t > 0\}$. We have $|\mathcal{T}| \geq 2$. We call this a two-or-more dosing plan. Consider the alternative dosing plan $\vec{b} = \left(\sqrt{\sum_{t\in\mathcal{T}}(d_t)^2},\ \underbrace{0,\ldots,0}_{N-1 \text{ times}}\right)$ built by reducing the dose in all sessions within the set $\{\mathcal{T} \setminus 1\}$ to 0 and increasing the dose in the first session. This is a single-dosage plan. Then,

$$
\begin{aligned}
C &= s\,(d_1 + d_2 + d_3 + \cdots + d_N) + s^2\rho\left((d_1)^2 + \cdots + (d_N)^2\right) \\
&= s\left(\sum_{t\in\mathcal{T}} d_t\right) + s^2\rho\left(\sum_{t\in\mathcal{T}}(d_t)^2\right) \qquad (\text{since } d_t = 0 \text{ when } t \notin \mathcal{T}) \\
&> s\left(\sqrt{\sum_{t\in\mathcal{T}}(d_t)^2}\right) + s^2\rho\left(\sum_{t\in\mathcal{T}}(d_t)^2\right). \qquad (2.15)
\end{aligned}
$$

The first equality holds because $\vec{d}$ is optimal and hence must make constraint (2.6) active. The subsequent strict inequality holds because $\sqrt{\sum_{t \in \mathcal{T}} (d_t)^2} < \sum_{t \in \mathcal{T}} d_t$ as $d_t > 0$ when $t \in \mathcal{T}$. This shows that the alternative plan is strictly feasible. Thus, there is an $\epsilon > 0$ such that the single-dosage plan $\vec{w} = \left(\sqrt{\sum_{t \in \mathcal{T}} (d_t)^2} + \epsilon, \underbrace{0, \ldots, 0}_{N-1 \text{ times}} \right)$ renders constraint (2.6) active. Moreover, let $\zeta > 0$ be such that $\sqrt{\sum_{t \in \mathcal{T}} (d_t)^2} + \zeta = \sum_{t \in \mathcal{T}} d_t$. We claim that $\zeta > \epsilon$. To see this, observe that

$$
\begin{aligned}
&s \sum_{t \in \mathcal{T}} d_t + s^2 \rho \sum_{t \in \mathcal{T}} (d_t)^2 \\
&\quad = s \left(\sqrt{\sum_{t \in \mathcal{T}} (d_t)^2} + \epsilon \right) + s^2 \rho \left(\sum_{t \in \mathcal{T}} (d_t)^2 + \epsilon^2 + 2\epsilon \sqrt{\sum_{t \in \mathcal{T}} (d_t)^2} \right),
\end{aligned}
$$

because both these quantities equal C. This is because both $\vec{d}$ and $\vec{w}$ render (2.6) active. Thus, canceling $\sum_{t \in \mathcal{T}} (d_t)^2$ from both sides and then substituting $\sum_{t \in \mathcal{T}} d_t = \sqrt{\sum_{t \in \mathcal{T}} (d_t)^2} + \zeta$ yields

$$
s \left(\sqrt{\sum_{t \in \mathcal{T}} (d_t)^2} + \zeta \right) = s \left(\sqrt{\sum_{t \in \mathcal{T}} (d_t)^2} + \epsilon \right) + s^2 \rho \left(\epsilon^2 + 2\epsilon \sqrt{\sum_{t \in \mathcal{T}} (d_t)^2} \right).
$$

Then, canceling $\sqrt{\sum_{t \in \mathcal{T}} (d_t)^2}$ from both sides and dividing through by $s > 0$ yields

$$
\zeta - \epsilon = s\rho \left(\epsilon^2 + 2\epsilon \sqrt{\sum_{t \in \mathcal{T}} (d_t)^2} \right).
$$

This implies that $\zeta - \epsilon > 0$ because every term on the right-hand side is positive. Now, the objective value of the original plan is

$$
\begin{aligned}
f_1(N, \vec{d}) &= \left(\alpha_0 - \frac{\beta_0}{s\rho} \right) \left(\sum_{t=1}^{N} d_t \right) + \frac{C\beta_0}{s^2 \rho} - \tau(N) \\
&\qquad \text{(algebra similar to (2.11))} \\
&= \left(\alpha_0 - \frac{\beta_0}{s\rho} \right) \left(\sum_{t \in \mathcal{T}} d_t \right) + \frac{C\beta_0}{s^2 \rho} - \tau(N) \qquad \text{(since } d_t = 0 \text{ when } t \notin \mathcal{T}\text{)}
\end{aligned}
$$

$$
\begin{aligned}
&= \left(\alpha_0 - \frac{\beta_0}{s\rho}\right)\left(\sqrt{\sum_{t\in\mathcal{T}}(d_t)^2 + \zeta}\right) + \frac{C\beta_0}{s^2\rho} - \tau(N) \\
&\qquad\qquad \left(\text{substituting } \sum_{t\in\mathcal{T}} d_t = \sqrt{\sum_{t\in\mathcal{T}}(d_t)^2 + \zeta}\right) \\
&< \left(\alpha_0 - \frac{\beta_0}{s\rho}\right)\left(\sqrt{\sum_{t\in\mathcal{T}}(d_t)^2 + \epsilon}\right) + \frac{C\beta_0}{s^2\rho} - \tau(N) \\
&\qquad\qquad \left(\epsilon < \zeta \text{ and } \alpha_0 - \tfrac{\beta_0}{s\rho} < 0\right) \\
&= f_1(N, \vec{w}). \qquad\qquad \text{(algebra similar to (2.11))}
\end{aligned}
$$

Thus, the alternative, single-dosage plan $\vec{w}$ is feasible and has an objective value strictly larger than the original optimal plan $\vec{d}$. This contradicts the optimality of the two-or-more dosing plan $\vec{d}$. Since $\vec{d}$ was an arbitrary two-or-more dosing plan, this demonstrates that no such plan can be optimal. Since at least one dose must be positive at optimality and since formula (2.9) provides the best value of a single dose, the proof is now complete.

2.3 Optimal Solution Using Karush–Kuhn–Tucker Conditions

The Karush–Kuhn–Tucker (KKT) necessary conditions for optimality in problem (P(N)) are given by

$$
-\begin{bmatrix} \alpha_0 + 2\beta_0 d_1 \\ \vdots \\ \alpha_0 + 2\beta_0 d_N \end{bmatrix} + \lambda s \begin{bmatrix} 1 + 2s\rho d_1 \\ \vdots \\ 1 + 2s\rho d_N \end{bmatrix} - \sum_{t=1}^{N} \mu_t \vec{e}_t = \vec{0} \tag{2.16}
$$

$$
(2.6) - (2.7) \tag{2.17}
$$

$$
\left(C - s\sum_{t=1}^{N} d_t - s^2\rho \sum_{t=1}^{N}(d_t)^2\right)\lambda = 0 \tag{2.18}
$$

$$
\mu_t d_t = 0,\ t = 1\colon N \tag{2.19}
$$

$$
\lambda \geq 0 \tag{2.20}
$$

$$
\mu_t \geq 0,\ t = 1\colon N. \tag{2.21}
$$

Here, $\vec{e}_t$ denotes an N-dimensional vector in which the tth entry is 1 and all other entries are 0, for $t = 1\colon N$. Lagrange multiplier λ is associated with the constraint (2.6), and Lagrange multipliers μ_t are associated with the constraint

$d_t \geq 0$, for $t = 1\colon N$. We first simplify these conditions as much as possible, without regard to the ordering of α_0/β_0 and $1/(s\rho)$. As such, these initial simplifications apply to the first two claims in the proposition.

For each $k = 1 : N$, let $\vec{d}(k)$ denote the N-dimensional vector

$$(d_1, \ldots, d_k, 0, 0, \ldots, 0),$$

where $d_1 > 0, d_2 > 0, \ldots, d_k > 0$. It suffices to consider such doses where the leading components are positive and tail components are 0 because both the objective function and constraint are symmetric with respect to permutations of components of doses. We use the above KKT conditions to characterize such doses.

For $\vec{d}(k)$, KKT condition (2.19) implies that $\mu_1 = \cdots = \mu_k = 0$. Thus, (2.16) yields

$$-\alpha_0 - 2\beta_0 d_t + \lambda s(1 + 2s\rho d_t) = 0,\ t = 1\colon k \tag{2.22}$$

$$-\alpha_0 + \lambda s = \mu_t,\ t = k + 1\colon N. \tag{2.23}$$

We rewrite (2.22) as

$$2d_t(-\beta_0 + \lambda s^2\rho) = \alpha_0 - \lambda s,\ t = 1\colon k. \tag{2.24}$$

Since the right-hand side of (2.24) does not depend on t, we obtain

$$d_1(-\beta_0 + \lambda s^2\rho) = \cdots = d_k(-\beta_0 + \lambda s^2\rho). \tag{2.25}$$

We prove by contradiction that $-\beta_0 + \lambda s^2\rho \neq 0$. This would imply from (2.25) that $d_1 = \cdots = d_k$. So suppose $\beta_0 = \lambda s^2\rho$. Substituting this into (2.24) yields $\alpha_0 = \lambda s$. Thus,

$$\frac{\alpha_0}{\beta_0} = \frac{\lambda s}{\lambda s^2\rho} = \frac{1}{\rho s} = \frac{\alpha}{s\beta}.$$

This contradicts the supposition that $(\alpha_0/\beta_0) \neq \alpha/(s\beta)$ in the first two claims of the proposition. This shows that $d_1 = \cdots = d_k$, which we denote by $d^\star(k) > 0$.

Equation (2.22) now implies that

$$\lambda = \frac{\alpha_0 + 2\beta_0 d^\star(k)}{s(1 + 2s\rho d^\star(k))} > 0, \tag{2.26}$$

and substituting this into (2.23) yields

$$\mu_t = \frac{2d^\star(k)(\beta_0 - \alpha_0 s\rho)}{1 + 2s\rho d^\star(k)},\ t = k + 1\colon N. \tag{2.27}$$

Now we consider the specific implications of the relative values of α_0/β_0 and $1/(s\rho)$, thereby leading to separate proofs of the first two claims in the proposition.

For the first claim, where $\alpha_0/\beta_0 > 1/(s\rho)$, the term $\beta_0 - \alpha_0 s\rho$ in the numerator of (2.27) is negative. Thus, no $d^\star(k) > 0$ can produce a $\mu_t \geq 0$, for $t = k+1 \colon N$, from (2.27). As such, for $k < N$, there are no solutions that satisfy the KKT condition (2.21). For $k = N$, (2.23) and hence (2.27) are moot. Thus, any positive dose that solves the quadratic equation

$$sNd + s^2\rho N d^2 = C \tag{2.28}$$

in d also satisfies the KKT condition (2.18) with the Lagrange multiplier λ derived in (2.26). In other words, administering the dose $d^\star(N)$ as given in formula (2.8) in each one of the N sessions is the only solution that satisfies all KKT conditions (2.16)–(2.21). Since problem (P(N)) does have a maximizer, this unique KKT solution must be it. This proves the first claim in the proposition.

For the second claim, where $\alpha_0/\beta_0 < 1/(s\rho)$, any $d^\star(k) > 0$ yields μ_t values from (2.27) that satisfy the KKT condition (2.21), for $t = k+1 \colon N$, when $k < N$. Furthermore, a $d^\star(k) > 0$ that solves

$$skd^\star(k) + s^2\rho k(d^\star(k))^2 = C, \tag{2.29}$$

satisfies the KKT condition (2.18) with the Lagrange multiplier λ derived in (2.26). The unique positive solution of this quadratic equation is given by

$$d^\star(k) = \frac{-1 + \sqrt{1 + \frac{4\rho C}{k}}}{2s\rho}. \tag{2.30}$$

Observe that this formula has the same form as (2.8) except that N is replaced by $k < N$. Finally, when $k = N$, (2.23) and hence (2.27) are moot. Nevertheless, any positive dose that solves the quadratic equation

$$sNd + s^2\rho N d^2 = C \tag{2.31}$$

in d satisfies the KKT condition (2.18) with the Lagrange multiplier λ derived in (2.26). Thus, administering the dose $d^\star(N)$ as given in formula (2.8) in each one of the N session, is the only solution that satisfies all KKT conditions (2.16)–(2.21). In summary, and combining everything from this case, we have N different solutions that satisfy all KKT conditions overall. These are given by $(\underbrace{d^\star(k), \ldots, d^\star(k)}_{k \text{ times}}, 0, \ldots, 0)$, where $d^\star(k)$ is available in formula (2.30), for $k = 1 \colon N$. In order to find the best among these N solutions, we compare their objective function values in problem (P(N)). The objective value of the kth solution equals $\alpha_0 k d^\star(k) + \beta_0 k (d^\star(k))^2 - \tau(N)$. Since the proliferation term $\tau(N)$ does not depend on k, it suffices to focus on the part $\alpha_0 k d^\star(k) + \beta_0 k (d^\star(k))^2$ without proliferation. It turns out that this expression is strictly decreasing in real numbers $k > 0$, if $\alpha_0/\beta_0 < 1/(s\rho)$. This can be proven by momentarily treating this expression as a function of real numbers $k > 0$ and then establishing that its derivative with respect to k is negative. We omit the proof here, because an identical argument will later be presented in detail, in a slightly different context, in (2.45)–(2.56) in Section 2.6.1. This implies that, among the N possible KKT solutions described above, the one where $k = 1$ actually produces the largest objective value. It must therefore be optimal. Observe that this solution precisely matches the one given by formula (2.9) in the proposition. This proves the second claim in the proposition.

2.4 Solution via Conversion into a Two-Variable Problem

We state and prove two intermediate lemmas.

Lemma 2.2 *Suppose x, y are positive real numbers. Let $N \geq 2$ be an integer.*

Claim 1 *Suppose $\sqrt{y} = x$. Then, there exist nonnegative real numbers $d_1, \ldots, d_N$ such that $x = \sum_{t=1}^N d_t$ and $y = \sum_{t=1}^N d_t^2$ if, and only if, exactly one among $d_1, \ldots, d_N$ is positive and all others are 0.*

Claim 2 *Suppose $x = \sqrt{Ny}$. Then, there exist nonnegative real numbers $d_1, \ldots, d_N$ such that $x = \sum_{t=1}^N d_t$ and $y = \sum_{t=1}^N d_t^2$ if, and only if, $d_1 = d_2 = \cdots = d_N$.*

Claim 3 *Suppose $\sqrt{y} < x < \sqrt{Ny}$. Then, there exist nonnegative real numbers $d_1, \ldots, d_N$ such that $x = \sum_{t=1}^N d_t$ and $y = \sum_{t=1}^N d_t^2$ if, and only if, at least two among $d_1, \ldots, d_N$ are positive and not all are equal.*

Proof We prove the three items separately.

Proof of Claim 1.
Suppose $\sqrt{y} = x$. Without loss of generality, let

$$d_1 = x,\ d_2 = d_3 = \cdots = d_N = 0. \tag{2.32}$$

Thus, $y = d_1^2 = x^2$. This constructively proves the “if” part. Now suppose that there are two distinct i and j such that $d_i > 0$ and $d_j > 0$. Then, letting

$x = d_i + d_j$ and $y = d_i^2 + d_j^2$ implies that $x^2 = d_i^2 + d_j^2 + 2d_i d_j > y$. Therefore, $x > \sqrt{y}$, thus contradicting the supposition that $x = \sqrt{y}$. This proves the "only if" part.

Proof of Claim 2.
Suppose $x = \sqrt{Ny}$. Then, setting

$$d_1 = \cdots = d_N = x/N \tag{2.33}$$

yields $y = \sum_{t=1}^{N} d_t^2 = \sum_{t=1}^{N} x^2/N^2 = x^2/N$. This constructively proves the "if" part. Now suppose that $d_1, \ldots, d_N$ are not equal. Then, setting $x = \sum_{t=1}^{N} d_t$ and $y = \sum_{t=1}^{N} d_t^2$ and then applying the Cauchy–Schwarz inequality (2.12) with $u_1 = \cdots = u_N = 1$ and $v_t = d_t$ yield $x < \sqrt{Ny}$. This contradicts the supposition that $x = \sqrt{Ny}$, thus proving the "only if" part.

Proof of Claim 3.
Suppose $\sqrt{y} < x < \sqrt{Ny}$. Let $2 \le k \le N$ be an integer such that $\sqrt{(k-1)y} < x \le \sqrt{ky}$. Let

$$d_1 = d_2 = \cdots = d_{k-1} = \frac{x - \sqrt{\frac{ky-x^2}{k-1}}}{k} \tag{2.34}$$

$$d_k = \frac{x + (k-1)\sqrt{\frac{ky-x^2}{k-1}}}{k} \tag{2.35}$$

$$d_{k+1} = d_{k+2} = \cdots = d_N = 0. \tag{2.36}$$

The term $ky - x^2$ inside the square root signs in (2.34) and (2.35) is nonnegative, by definition of k. Thus, the square root is a real number. Moreover, the expressions on the right-hand sides of (2.34) and (2.35) are distinct if $ky - x^2 > 0$. Then, since $k \ge 2$, expressions (2.34) and (2.35) include at least two doses that are unequal. If, on the other hand, $ky = x^2$, then $k < N$ because we know that $x^2 < Ny$. In this case, although the expressions (2.34) and (2.35) are identical, we have at least one dose that is 0 from (2.36). In addition, by definition of k, we have

$$(k-1)y < x^2 \Rightarrow ky - x^2 < y \Rightarrow \frac{ky - x^2}{k-1} < \frac{y}{k-1}$$

$$\Rightarrow \sqrt{\frac{ky - x^2}{k-1}} < \sqrt{\frac{y}{k-1}}. \tag{2.37}$$

Similarly,

$$(k-1)y < x^2 \Rightarrow \frac{y}{k-1} < \frac{x^2}{(k-1)^2} \Rightarrow \sqrt{\frac{y}{k-1}} < \frac{x}{k-1}. \tag{2.38}$$

Inequalities (2.37) and (2.38) imply that $\sqrt{\frac{ky-x^2}{k-1}} < \frac{x}{k-1} \leq x$. Here, the last inequality holds because $k \geq 2$. Thus, the dose in (2.34) is positive. The dose in (2.35) is also positive, for instance, because x is positive.

Observe that

$$\begin{aligned}\sum_{t=1}^{N} d_t &= (k-1)\frac{x - \sqrt{\frac{ky-x^2}{k-1}}}{k} + \frac{x + (k-1)\sqrt{\frac{ky-x^2}{k-1}}}{k} \\ &= \frac{(k-1)x - (k-1)\sqrt{\frac{ky-x^2}{k-1}} + x + (k-1)\sqrt{\frac{ky-x^2}{k-1}}}{k} = x.\end{aligned}$$

Similarly,

$$\begin{aligned}\sum_{t=1}^{N} d_t^2 &= (k-1)\left(\frac{x - \sqrt{\frac{ky-x^2}{k-1}}}{k}\right)^2 + \left(\frac{x + (k-1)\sqrt{\frac{ky-x^2}{k-1}}}{k}\right)^2 \\ &= (k-1)\left(\frac{x^2 - 2x\sqrt{\frac{ky-x^2}{k-1}} + \frac{ky-x^2}{k-1}}{k^2}\right) \\ &\quad + \left(\frac{x^2 + 2x(k-1)\sqrt{\frac{ky-x^2}{k-1}} + (k-1)^2\frac{ky-x^2}{k-1}}{k^2}\right) \\ &= \frac{(k-1)x^2 + (k-1)\frac{ky-x^2}{k-1} + x^2 + (k-1)^2\frac{ky-x^2}{k-1}}{k^2} \\ &= \frac{kx^2 + (k-1)\frac{ky-x^2}{k-1}(1 + (k-1))}{k^2} \\ &= \frac{kx^2 + k^2y - kx^2}{k^2} = y.\end{aligned}$$

Thus, we have constructively established the "if" part. Conversely, suppose either that only one among $d_1, \ldots, d_N$ is positive (say it is d_1 without loss of generality) or that $d_1 = \cdots = d_N$. In the former case, if we set $x = d_1$ and $y = d_1^2$, then clearly $x = \sqrt{y}$. In the latter case, if we set $x = Nd_1$ and $y = Nd_1^2$, then $x = \sqrt{Ny}$. Both these contradict the supposition that $\sqrt{y} < x < \sqrt{Ny}$. This proves the "only if" part. □

Lemma 2.3 *The* N*-variable problem* $(P(N))$ *is equivalent to the 2-variable problem*

$$(2VAR1(N))\ f_1^\star(N) = \max_{x,y}\ \alpha_0 x + \beta_0 y - \tau(N) \tag{2.39}$$

$$sx + s^2\rho y \leq C \tag{2.40}$$

$$\sqrt{y} \leq x \tag{2.41}$$

$$x \leq \sqrt{Ny} \tag{2.42}$$

$$x \geq 0, y \geq 0 \tag{2.43}$$

in the sense summarized in the three claims below.

Claim 1 *If* $\vec{d}$ *is a feasible solution to problem* $(P(N))$*, then the solution* $x = \sum_{t=1}^{N} d_t$ *and* $y = \sum_{t=1}^{N} d_t^2$ *is feasible to problem* $(2VAR1(N))$*; solutions* $\vec{d}$ *and* (x, y) *have identical objective function values in the two problems.*

Claim 2 *Conversely, if* (x, y) *is feasible to problem* $(2VAR1(N))$*, then Lemma 2.2 provides a method to construct a feasible solution* $\vec{d}$ *to problem* $(P(N))$ *such that the two solutions have identical objective function values in the two problems.*

Claim 3 *An optimal solution to one problem corresponds to an optimal solution to the other problem, and the two optimal objective values are equal.*

Proof For the first claim, suppose $\vec{d}$ is a feasible solution to problem (P(N)). That is, it satisfies constraints (2.6) and (2.7). Let $x = \sum_{t=1}^{N} d_t$ and $y = \sum_{t=1}^{N} d_t^2$. Then, it is clear that the pair (x, y) is feasible to constraints (2.40), (2.43), and that the objective value of $\vec{d}$ in (2.5) is equal to the objective value of (x, y) in (2.39). Moreover,

$$x^2 = (d_1 + \cdots + d_N)^2 = d_1^2 + \cdots + d_N^2 + 2\sum_{i=1}^{N}\sum_{\substack{j=1 \\ j\neq i}}^{N}(d_i d_j)$$

$$= y + 2\sum_{i=1}^{N}\sum_{\substack{j=1 \\ j\neq i}}^{N}(d_i d_j) \geq y.$$

Here, the inequality follows because both d_i and d_j are nonnegative owing to constraint (2.7). Consequently, (2.41) holds. In addition, applying the Cauchy-Schwarz inequality (2.12) with $u_1 = \cdots = u_N = 1$ and $v_t = d_t$, for $t = 1\colon N$, we see that $x = \sum_{t=1}^{N} d_t \leq \sqrt{N}\sqrt{\sum_{t=1}^{N} d_t^2} = \sqrt{Ny}$. That is, (2.42) holds. This establishes that the pair (x, y) is feasible to problem (2VAR1(N)). This proves the first claim.

For the second claim, suppose the pair (x, y) is feasible to problem (2VAR1(N)). We consider three cases with respect to constraints (2.41)–(2.42): (i) $\sqrt{y} = x$, (ii) $x = \sqrt{Ny}$, and (iii) $\sqrt{y} < x < \sqrt{Ny}$. In each case, Lemma 2.2 provides a $\vec{d} \geq 0$ such that $x = \sum_{t=1}^{N} d_t$ and $y = \sum_{t=1}^{N} d_t^2$. Since (x, y) satisfies constraint (2.40), this implies that $\vec{d}$ satisfies constraint (2.6). Similarly, the objective function values of (x, y) in (2VAR1(N)) and $\vec{d}$ in (P(N)) are equal, thus establishing the second claim.

The third claim follows immediately from the first two claims. □

We are now ready to complete the proof of Proposition 2.1 with the help of Figure 2.1, which illustrates the feasible region for problem (2VAR1(N)).

Refer to Figure 2.1. Since the coefficients of both x and y in the linear objective function (2.39) are positive, we wish to push its contour lines as much outward in the positive $x - y$ quadrant as possible. Thus, optimal solutions only occur on the line segment AB, which is the boundary of the linear constraint (2.40).

Specifically, if $\frac{\alpha_0}{\beta_0} > \frac{1}{s}\frac{\alpha}{\beta}$, then the slope of the objective function contour lines (dotted in the figure) is steeper than the slope of line segment AB.

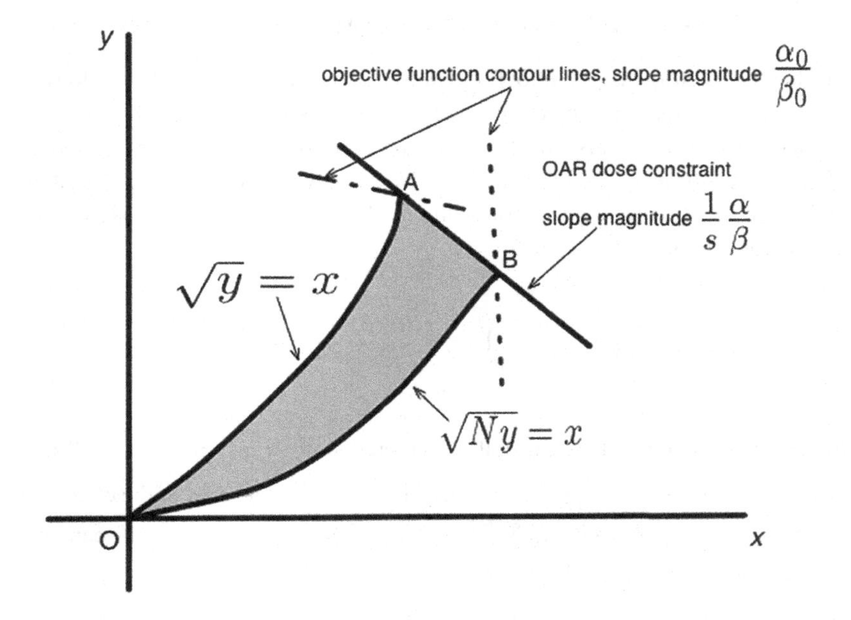

Figure 2.1 Feasible region (shaded in gray) and objective function contour lines for problem (2VAR1(N)). Reused with permission from [61, figure 4].

The unique optimal solution then occurs at point B, where $x > 0, y > 0$ and $x = \sqrt{Ny}$. Thus, from Claim 2 in Lemma 2.2, the corresponding optimal solution to problem (P(N)) must have $d_1 = d_2 = \cdots = d_N$. We denote all these doses by $d^\star(N)$. Since constraint (2.6) must be active at an optimal solution, we solve the quadratic equation $Nsd^\star(N) + Ns^2\rho(d^\star(N))^2 = C$ to find a positive value of $d^\star(N)$. This yields formula (2.8).

Similarly, if $\frac{\alpha_0}{\beta_0} < \frac{1}{s}\frac{\alpha}{\beta}$, then the slope of the objective function contour lines (dash-dotted in the figure) is flatter than the slope of line segment AB. The unique optimal solution then occurs at point A, where $x > 0, y > 0$, and $x = \sqrt{y}$. Thus, from Claim 1 in Lemma 2.2, the corresponding optimal solution to problem (P(N)) is such that $d_1 > 0$ and $d_2 = \cdots = d_N = 0$. We denote d_1 by d°. Since constraint (2.6) must be active at an optimal solution, we solve the quadratic equation $sd^\circ + s^2\rho(d^\circ)^2 = C$ to find d°. This yields formula (2.9).

Finally, if $\frac{\alpha_0}{\beta_0} = \frac{1}{s}\frac{\alpha}{\beta}$, then the slope of the objective function contour lines (not shown in the figure) is equal to the slope of line segment AB. Then, every point on line segment AB is optimal. This establishes the third claim.

2.5 Geometric Derivation of Optimal Solution

Further geometric insight into Proposition 2.1 can be obtained by plotting the feasible region and objective function contours for problem (P(N)). The discussion below can also be seen as a perhaps slightly less rigorous proof of the proposition itself. Observe that, by completing squares appropriately and recalling that $\rho = \beta/\alpha$, constraint (2.6) can be expressed as the N-dimensional ball

$$\sum_{t=1}^{N}\left(d_t + \frac{\alpha/\beta}{2s}\right)^2 \leq \frac{C(\alpha/\beta)}{s^2} + \frac{N(\alpha/\beta)^2}{4s^2}.$$

In the $\vec{d}$ coordinate system, this ball is centered at $d_1 = \cdots = d_N = -\frac{\alpha/\beta}{2s}$ and has radius $\sqrt{\frac{C(\alpha/\beta)}{s^2} + \frac{N(\alpha/\beta)^2}{4s^2}}$. Similarly, the contour at level Z of the objective function, which is given by $\alpha_0\sum_{t=1}^{N} d_t + \beta_0\sum_{t=1}^{N}(d_t)^2 - \tau(N) = Z$, can also be expressed as the N-dimensional hypersphere

$$\sum_{t=1}^{N}\left(d_t + \frac{(\alpha_0/\beta_0)}{2}\right)^2 = \frac{2\beta_0 Z + 2\tau(N) + (\alpha_0)^2}{2(\beta_0)^2}.$$

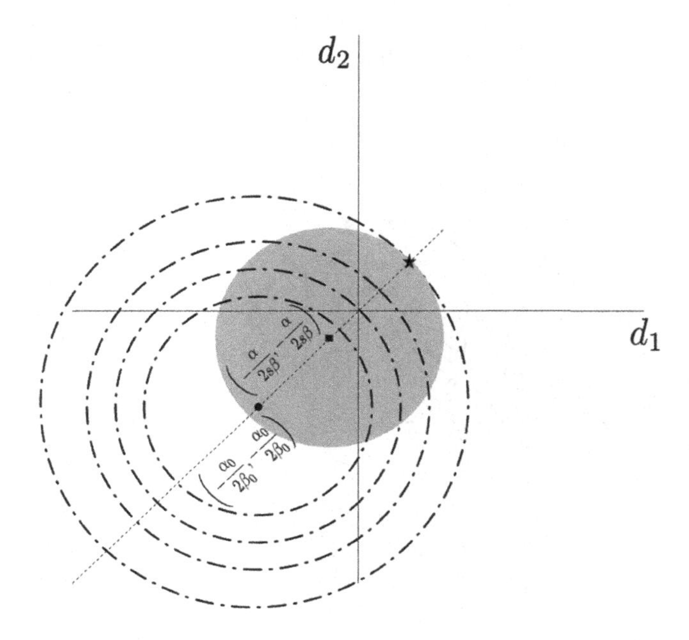

Figure 2.2 Schematic illustration of Claim 1 in Proposition 2.1, where equal dosage is optimal. The feasible region equals the intersection of the shaded gray circle with the nonnegative quadrant in $\mathbb{R}^2$. The dash-dotted circles represent contours of the objective function. Reused with permission from [61, figure 5].

This hypersphere is centered at $d_1 = \cdots = d_N = -\frac{\alpha_0/\beta_0}{2}$. Figures 2.2 and 2.3 illustrate this for the special case of $N = 2$. In these figures, the feasible region equals the intersection of the nonnegative quadrant in the (d_1, d_2) plane with a disc centered at $(-\frac{\alpha}{2s\beta}, -\frac{\alpha}{2s\beta})$. Similarly, the contour of the objective function at level Z is a circle centered at $(-\frac{\alpha_0}{2\beta_0}, -\frac{\alpha_0}{2\beta_0})$. Figure 2.2 shows that an equal-dosage solution $d_1 = d_2$ (black star) is optimal when the center of the objective function contour (black circle) is farther from the origin than the center of the feasible region disc (black square). This is the first claim in the proposition. Figure 2.3 shows that single-dosage solutions $d_1 > 0, d_2 = 0$ or $d_1 = 0, d_2 > 0$ (black stars) are optimal when the center of the objective function contour (black circle) is closer to the origin than the center of the feasible region disc (black square). This is the second claim in the proposition. The constraint disc and the optimal contour circle coincide in the third claim in the proposition, and hence any feasible dosing plan on the outer boundary of the disc is optimal. This case is not illustrated.

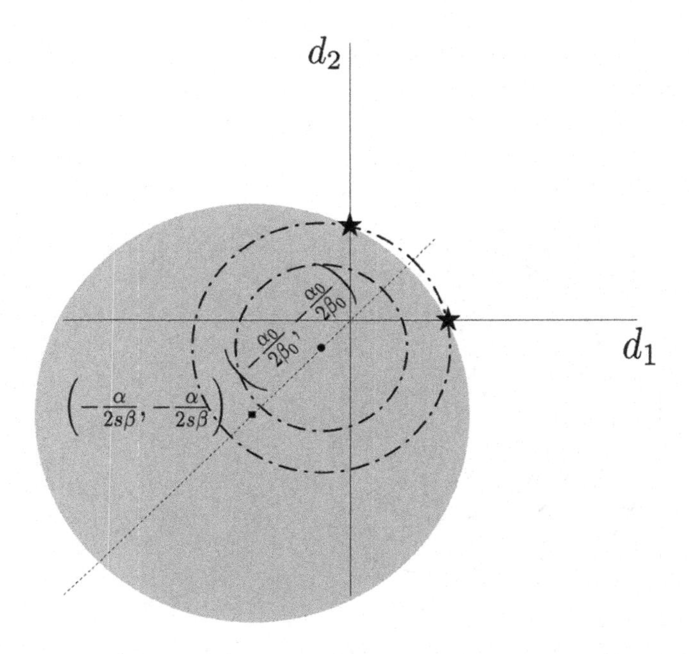

Figure 2.3 Schematic illustration of Claim 2 in Proposition 2.1, where single dosage is optimal. The feasible region equals the intersection of the shaded gray circle with the nonnegative quadrant in $\mathbb{R}^2$. The dash-dotted circles represent contours of the objective function. Reused with permission from [61, figure 5].

2.6 Optimal Number of Sessions

Proposition 2.1 establishes that it is optimal to administer dose (2.9) in a single session and not to administer any dose in other sessions, when $\frac{\alpha_0}{\beta_0} \leq \frac{\alpha/\beta}{s}$. This holds in problem (P(N)) regardless of N. Thus, the optimal number of sessions equals 1 with a dose given by (2.9), when $\frac{\alpha_0}{\beta_0} \leq \frac{\alpha/\beta}{s}$. This section therefore only focuses on the case of $\frac{\alpha_0}{\beta_0} > \frac{\alpha/\beta}{s}$.

Theorem 2.4 *Suppose* $\frac{\alpha_0}{\beta_0} > \frac{(\alpha/\beta)}{s}$. *Let* $r = 1/(2s\rho)$ *and* $\eta = \ln(2)/T_{double}$ *for brevity. Let*

$$N^\star = \frac{4\rho C}{\left(\frac{\eta+\sqrt{\eta^2+2\eta r(\alpha_0-\alpha\beta_0/(s\beta))}}{r(\alpha_0-\alpha\beta_0/(s\beta))}+1\right)^2-1}. \tag{2.44}$$

There are three possibilities.

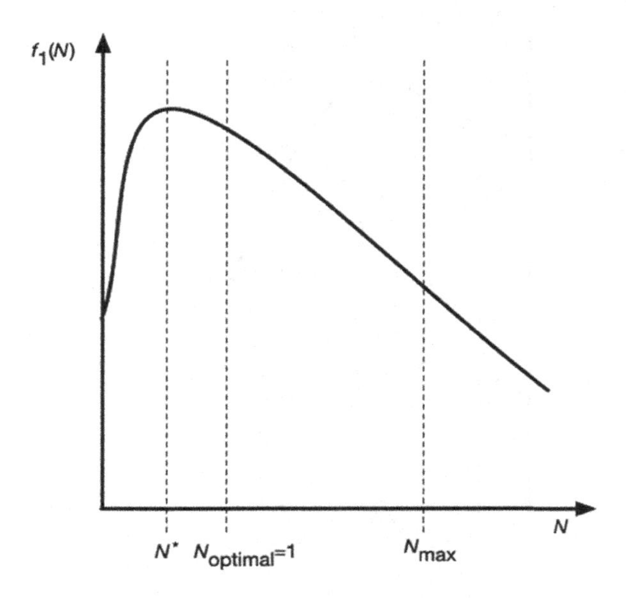

Figure 2.4 Schematic illustration of the first case in Theorem 2.4, where $N^\star < 1$.

1. *if* $N^\star < 1$, *then it is optimal to administer* 1 *session (Figure 2.4);*
2. *if* $1 \leq N^\star \leq N_{max}$, *then it is optimal to administer either* $\lfloor N^\star \rfloor$ *or* $\lceil N^\star \rceil$ *sessions depending on which one of these two integers maximizes the objective function* $f_1^\star(N)$ *in (P) (Figure 2.5); and*
3. *if* $N^\star > N_{max}$, *then it is optimal to administer* N_{max} *sessions (Figure 2.6).*

We denote the optimal number of treatment sessions by $N_{optimal}$ *and the corresponding optimal dose per session by* $d^\star(N_{optimal})$ *as defined via* (2.8).

2.6.1 Solution Using Calculus

We substitute the dose per session (2.8), which is optimal in the case $\frac{\alpha_0}{\beta_0} > \frac{\alpha/\beta}{s}$, into the objective function of problem (P). This allows us to rewrite problem (P) as

$$f_1^\star = \max_{\substack{1 \leq N \leq N_{\max} \\ N \text{ integer}}} f_1^\star(N) = \alpha_0 N d^\star(N) + \beta_0 N (d^\star(N))^2 - \tau(N). \qquad (2.45)$$

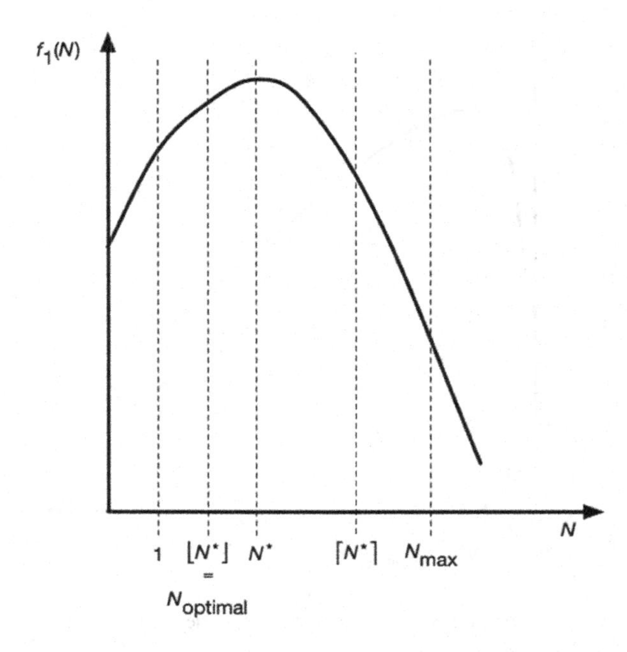

Figure 2.5 Schematic illustration of the second case in Theorem 2.4, where $1 \leq N^\star \leq N_{\max}$.

We introduce

$$D^\star(N) = Nd^\star(N) = N\left(\frac{-1+\sqrt{1+4\rho C/N}}{2s\rho}\right)$$
$$= \left(\frac{-N+\sqrt{N^2+4\rho CN}}{2s\rho}\right) \tag{2.46}$$

as the total dose administered in N sessions, and thereby express $f_1^\star(N)$ from (2.45) as

$$f_1^\star(N) = \alpha_0 D^\star(N) + \beta_0 \frac{(D^\star(N))^2}{N} - \frac{(N-1)\ln(2)}{T_{\text{double}}}. \tag{2.47}$$

We temporarily view $f_1^\star(N)$ as a function of real numbers $N > 0$ and study its behavior using calculus. We have

$$\frac{df_1^\star(N)}{dN} = \alpha_0 \frac{dD^\star(N)}{dN} + \beta_0 \frac{2ND^\star(N)\frac{dD^\star(N)}{dN} - (D^\star(N))^2}{N^2} - \frac{\ln(2)}{T_{\text{double}}}$$
$$= \alpha_0 \frac{dD^\star(N)}{dN} + 2\beta_0 d^\star(N)\frac{dD^\star(N)}{dN} - \beta_0(d^\star(N))^2 - \frac{\ln(2)}{T_{\text{double}}}. \tag{2.48}$$

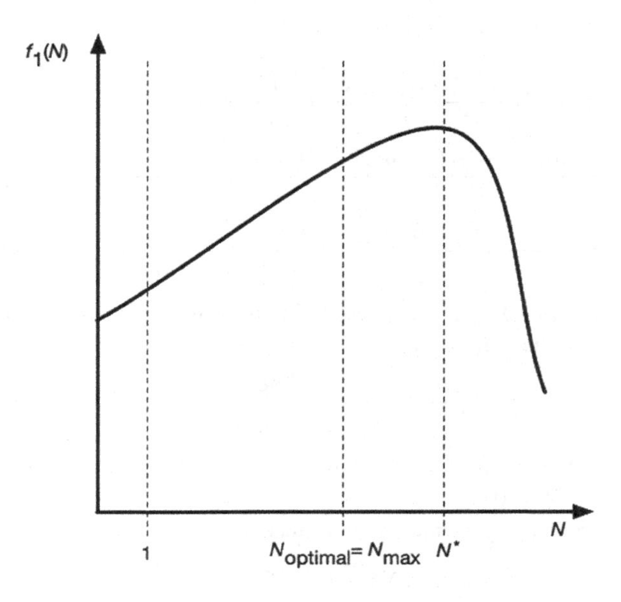

Figure 2.6 Schematic illustration of the third case in Theorem 2.4, where $N^{\star} > N_{\max}$.

Utilizing (2.46) we get

$$\frac{dD^{\star}(N)}{dN} = \frac{1}{2s\rho}\left(-1 + \frac{2N + 4\rho C}{2\sqrt{N^2 + 4\rho CN}}\right)$$
$$= \frac{1}{2s\rho}\frac{\left(\sqrt{1 + 4\rho C/N} - 1\right)^2}{2\sqrt{1 + 4\rho C/N}}. \tag{2.49}$$

For brevity, we let

$$\chi(N) = \sqrt{1 + 4\rho C/N} - 1. \tag{2.50}$$

We have $d^{\star}(N) = r\chi(N)$ and $dD^{\star}(N)/dN = r\chi^2(N)/(2(\chi(N) + 1))$. Substituting these into (2.48) we obtain

$$\frac{df_1^{\star}(N)}{dN} = \alpha_0\frac{r\chi^2(N)}{2(\chi(N) + 1)} + 2\beta_0\frac{r^2\chi^3(N)}{2(\chi(N) + 1)} - \beta_0 r^2\chi^2(N) - \frac{\ln(2)}{T_{\text{double}}}$$
$$= \frac{(\alpha_0 - 2\beta_0 r)r\chi^2(N)}{2(\chi(N) + 1)} - \frac{\ln(2)}{T_{\text{double}}} \tag{2.51}$$
$$= \frac{(\alpha_0 - \alpha\beta_0/(s\beta))r\chi^2(N)}{2(\chi(N) + 1)} - \frac{\ln(2)}{T_{\text{double}}}. \tag{2.52}$$

We now have

$$\frac{d^2 f_1^\star(N)}{dN^2} = \frac{d}{dN}\left(\frac{df_1^\star(N)}{dN}\right) = \frac{d}{d\chi}\left(\frac{df_1^\star(N)}{dN}\right)\frac{d\chi(N)}{dN} \tag{2.53}$$

$$= \frac{(\alpha_0 - \alpha\beta_0/(s\beta))r}{2}\frac{2\chi(N)(\chi(N)+1) - \chi^2(N)}{(\chi(N)+1)^2}\frac{d\chi(N)}{dN} \tag{2.54}$$

$$= \frac{(\alpha_0 - \alpha\beta_0/(s\beta))r}{2}\frac{\chi^2(N)+2\chi(N)}{(\chi(N)+1)^2}\frac{d\chi(N)}{dN}. \tag{2.55}$$

Now recall that $\chi(N) = \sqrt{1+4C\rho/N} - 1$ and hence $\chi(N) > 0$ for all $N > 0$. Moreover, from the definition in (2.50), we obtain

$$\frac{d\chi(N)}{dN} = \frac{-2\rho C}{N^2\sqrt{1+4\rho C/N}} < 0,$$

for all $N > 0$. Therefore, $d^2 f_1^\star(N)/dN^2 < 0$ and hence $f_1^\star(N)$ is strictly concave for all $N > 0$.

We now prove two intermediate lemmas.

Lemma 2.5 *We have that $D^\star(N) \le C/s$, for all $N \ge 1$.*

Proof Observe from (2.46) that

$$D^\star(N) = \frac{-N+\sqrt{N^2+4\rho CN}}{2s\rho} \le \frac{C}{s} \Leftrightarrow -N + \sqrt{N^2+4\rho CN} \le 2\rho C$$
$$\Leftrightarrow \sqrt{N^2+4\rho CN} \le N + 2\rho C \Leftrightarrow N^2 + 4\rho CN \le (N+2\rho C)^2$$
$$\Leftrightarrow 0 \le 4\rho^2 C^2.$$

This establishes the claim. □

Lemma 2.6 *Suppose $\frac{\alpha_0}{\beta_0} > \frac{(\alpha/\beta)}{s}$. Then, there exists an integer $\hat{N} \ge 2$ such that $f_1^\star(\hat{N}) < f_1^\star(\hat{N}-1)$. That is, the objective in* (2.45) *eventually decreases.*

Proof We introduce $z(N) = \alpha_0 N d^\star(N) + \beta_0 N(d^\star(N))^2$ as the part of $f_1^\star(N)$ without the proliferation term $\tau(N)$. A close inspection of the derivative of $f_1^\star(N)$ from (2.51) reveals that

$$\frac{dz(N)}{dN} = \frac{(\alpha_0 - \alpha\beta_0/(s\beta))r\chi^2(N)}{2(\chi(N)+1)} > 0. \tag{2.56}$$

Here, the strict inequality holds because $(\alpha_0 - \alpha\beta_0/(s\beta)) > 0$ by assumption, and $r, \chi(N)$ are both positive. Thus, $z(N)$ can be viewed as a strictly increasing sequence indexed by $N \ge 1$. Lemma 2.5 implies that

$$z(N) = \alpha_0 N d^\star(N) + \beta_0 N(d^\star(N))^2 \le \alpha_0 N d^\star(N) + \beta_0 N^2(d^\star(N))^2$$
$$= \alpha_0 D^\star(N) + \beta_0(D^\star(N))^2 \le \alpha_0\left(\frac{C}{s}\right) + \beta_0\left(\frac{C}{s}\right)^2, \ \forall N \ge 1.$$

Thus, $z(N)$ is an increasing sequence that is bounded above. It therefore converges. Specifically, there exists a positive integer $\bar{N} \geq 2$ such that $z(N) - z(N-1) < \ln(2)/T_{\text{double}} = \tau(N) - \tau(N-1)$ for all $N \geq \bar{N}$. Thus, by moving terms around the inequality, we obtain $f_1^\star(N) = z(N) - \tau(N) < z(N-1) - \tau(N-1) = f_1^\star(N-1)$, for all $N \geq \bar{N}$. □

Remark 2.7 *The calculation in* (2.56) *shows that the larger the number of sessions, the better, if* $\frac{\alpha_0}{\beta_0} > \frac{(\alpha/\beta)}{s}$ *and if the tumor does not proliferate.*

Since $f_1^\star(N)$ is strictly concave as a function of real numbers $N > 0$ and it eventually decreases by Lemma 2.6, $f_1^\star(N)$ must have a maximizer. We equate the derivative of $f_1^\star(N)$ from (2.51) to 0 in an attempt to find this maximizer. We get the quadratic equation

$$r(\alpha_0 - \alpha\beta_0/(s\beta))\chi^2(N) - 2\eta\chi(N) - 2\eta = 0 \tag{2.57}$$

in $\chi(N)$. Since $\chi(N)$ must be positive, we investigate whether (2.57) has a positive solution. Solutions of this equation are

$$\chi(N) = \frac{\eta \pm \sqrt{\eta^2 + 2\eta r_1(\alpha_0 - \alpha\beta_0/(s\beta))}}{r_1(\alpha_0 - \alpha_1\beta_0/\sigma_1\beta_1)}.$$

Since $\frac{\alpha_0}{\beta_0} > \frac{(\alpha_1/\beta_1)}{s}$, the only positive solution is

$$\chi^\star(N) = \frac{\eta + \sqrt{\eta^2 + 2\eta r(\alpha_0 - \alpha\beta_0/(s\beta))}}{r(\alpha_0 - \alpha\beta_0/(s\beta))}. \tag{2.58}$$

Moreover, since $\chi^\star(N) = \sqrt{1 + 4\rho C/N} - 1$, formula (2.58) yields (2.44).

The three claims in the theorem then follow, since $f_1^\star(N)$ is strictly concave and the optimal number of sessions must be an integer between 1 and N_{max}.

2.7 Numerical Experiments

Table 2.1 lists N_{optimal} and $d^\star(N_{\text{optimal}})$ values for various $T_{\text{double}}, \rho, s$ triples. Tumor parameters were fixed at $\alpha_0 = 0.35$ Gy^{-1} and $\beta_0 = 0.035$ Gy^{-2}, for an α_0/β_0 ratio of 10 Gy. The conventional number of treatment sessions was set to $T = 35$ with a dose of $\delta = 1.2857$ Gy per session. These numbers are typical of head-and-neck cancer with spinal cord as the OAR. The maximum possible number of sessions, N_{max}, was set at 100. Table 2.2 repeats these calculations with tumor $\alpha_0 = 0.7$ Gy^{-1}, for an α_0/β_0 ratio of 20 Gy.

Formula (2.44) shows that $N^\star$ is decreasing in η and hence increasing in T_{double}. Since the transformation between $N^\star$ and N_{optimal} is monotone, N_{optimal} is also increasing in T_{double}. This is intuitive because the larger the doubling time, the slower the tumor grows, and hence longer treatment

Table 2.1. $(N_{optimal}, d^{\star}(N_{optimal})$ *Gy) for different* T_{double}, ρ, s *values with* α_0/β_0 *ratio of* 10 *Gy.*

	$\rho = 1/5\ \mathrm{Gy}^{-1}$			$\rho = 1/4\ \mathrm{Gy}^{-1}$			$\rho = 1/3\ \mathrm{Gy}^{-1}$		
	T_{double} (days)			T_{double} (days)			T_{double} (days)		
s	2	3	4	2	3	4	2	3	4
1	(5,5.43)	(7,4.33)	(10,3.38)	(6,4.61)	(9,3.52)	(12,2.88)	(7,3.96)	(11,2.95)	(16,2.28)
0.8	(4,7.84)	(7,5.41)	(9,4.55)	(6,5.76)	(10,4.09)	(13,3.40)	(8,4.54)	(13,3.29)	(17,2.73)
0.6	(2,16.09)	(4,10.45)	(5,9.04)	(5,8.64)	(8,6.35)	(11,5.10)	(9,5.61)	(14,4.17)	(19,3.37)

Table 2.2. ($N_{optimal}, d^{\star}(N_{optimal})$ *Gy) for different* T_{double}, ρ, s *values with* α_0/β_0 *ratio of* 20 *Gy.*

	$\rho = 1/5\ \mathrm{Gy}^{-1}$			$\rho = 1/4\ \mathrm{Gy}^{-1}$			$\rho = 1/3\ \mathrm{Gy}^{-1}$		
	T_{double} (days)			T_{double} (days)			T_{double} (days)		
s	2	3	4	2	3	4	2	3	4
1	(14,2.64)	(20,2.02)	(26,1.64)	(16,2.34)	(23,1.79)	(30,1.45)	(19,2.02)	(28,1.52)	(36,1.26)
0.8	(16,2.99)	(23,2.26)	(28,1.93)	(19,2.58)	(27,1.97)	(34,1.64)	(22,2.27)	(32,1.72)	(41,1.42)
0.6	(18,3.64)	(25,2.82)	(31,2.37)	(22,3.08)	(31,2.36)	(38,2.01)	(27,2.61)	(39,1.97)	(49,1.65)

sessions are optimal. Since the transformation between $N^{\star}$ and N_{optimal} is monotone, this implies that N_{optimal} is decreasing. Table 2.1 confirms this trend.

The trend in $N^{\star}$ (and hence in N_{optimal}) as a function of s is not evident from formula (2.44) because s appears in multiple terms in a complicated manner in the denominator. Indeed, Table 2.1 demonstrates that N_{optimal} can decrease or increase as s decreases, for each fixed ρ, T_{double} pair.

The trend in $N^{\star}$ (and hence in N_{optimal}) as a function of ρ is not evident from formula (2.44) because ρ appears in multiple terms in a complicated manner in the numerator and denominator. Nevertheless, N_{optimal} appears to increase as ρ increases, for each fixed s, T_{double} pair in Table 2.1. It may be possible to prove this by checking whether the derivative of $N^{\star}$ with respect to ρ is positive. We do not attempt this here, as the derivative is rather tedious to compute.

A cell-by-cell comparison of Tables 2.1 and 2.2 demonstrates that N_{optimal} is increasing as α_0 increases when other parameters are fixed. We prove this next and also show that the effect of β_0 is the opposite.

Lemma 2.8 *When all other parameters are fixed, $N^{\star}$ is increasing (and hence $N_{optimal}$ is increasing) in α_0. On the other hand, when all other parameters are fixed, $N^{\star}$ is decreasing (and hence $N_{optimal}$ is decreasing) in β_0 as long as $\beta_0 < \alpha_0 s\beta/\alpha$.*

Proof Since $N^{\star}$ is derived from (2.58) using the defining relationship (2.50), we know that $N^{\star}$ increases as $\chi^{\star}(N)$ decreases. Thus, it suffices to prove that $\chi^{\star}(N)$ in (2.58) decreases as α_0 increases. Finally, this property holds because each term in $\chi^{\star}(N)$ decreases as α_0 increases.

The proof for β_0 is similar, except that the negative sign in front of the β_0 term in (2.58) renders the trend in sensitivity to β_0 opposite to that of α_0. Since formula (2.44) is only valid when $\beta_0 < \alpha_0 s\beta/\alpha$, we include this restriction in the hypothesis of the lemma. □

Equipped with the analytical tools and insights from the chapter, we will generalize to the case of multiple OAR next.

Bibliographic Notes

A detailed introduction to the Cauchy–Schwarz inequality is available in [114]. Early formulations of the optimal fractionation problem based on the LQ dose-response framework with a single OAR can be traced back to [45, 46, 47, 48, 49, 50, 51, 52, 53], and [11, 72]. The notion of sparing factors was employed in [88, 103, 123]. Details about how to compute these sparing factors were included in [103]. Geometric insights into the fractionation problem using hypersphere-shaped objective function contours and a ball-shaped feasible region were provided in [88]. The observation in Remark 2.7 was made in [75, 103, 123]. Strict concavity of the objective function $f_1^{\star}(N)$ as in expression (2.45) was established in [22, 103]. A brief overview of the results in this chapter was included, without complete proofs, in a tutorial [61]. All results in this chapter can be extended to a model wherein tumor proliferation begins after an initial lag of T_{lag} days. In that model, $\tau(N) = \frac{(N-1-T_{\text{lag}})^+ \ln(2)}{T_{\text{double}}}$. Here, $x^+ = \max\{0, x\}$ for any real number x. This extension only requires minor changes to the proofs provided in this chapter. The equivalence between problem (P(N)) and the two-variable problem (2VAR1(N)) as in Lemma 2.3 was noted in [102]. The papers by Saberian et al. cited above are based on the doctoral dissertation [101]. See [50, 51, 95, 132], for instance, for head-and-neck or lung cancer data where the condition in the first claim of Proposition 2.1 holds. Similarly, see [39, 53, 127], for discussions related to the condition in the second claim of Proposition 2.1. Examples of clinical trials with very short treatment courses for prostate cancer include [15, 25, 29, 63, 131, 138, 139, 140].

Exercises

Exercise 2.9 *Either prove or disprove by counterexample that $N^{\star}$ defined in (2.44) is increasing in $\rho > 0$.

Exercise 2.10 Restate, as appropriate, all theoretical results and their derivations in Section 2.6 for the case where the tumor proliferation term is given by $\tau(N) = \frac{(N-1-T_{\text{lag}})^+ \ln(2)}{T_{\text{double}}}$ as described in the Bibliographic Notes above.

3

Fractionation with Multiple Organs-at-Risk

In this chapter, we will study a generalization that includes multiple OAR into problem (P) from Chapter 2. We use $\mathcal{M} = \{1, \ldots, M\}$ to denote the set of $M \geq 1$ OAR. The optimal fractionation problem is now formulated as

$$(\text{Q})\ f^\star = \max_{\vec{d},\ N}\ \alpha_0 \sum_{t=1}^{N} d_t + \beta_0 \sum_{t=1}^{N} (d_t)^2 - \tau(N) \tag{3.1}$$

$$s_m \sum_{t=1}^{N} d_t + s_m^2 \rho_m \sum_{t=1}^{N} (d_t)^2 \leq \underbrace{T_m \delta_m (1 + \rho_m \delta_m)}_{C_m},\ m \in \mathcal{M} \tag{3.2}$$

$$\vec{d} \geq 0 \tag{3.3}$$

$$1 \leq N \leq N_{\max},\ \text{integer}. \tag{3.4}$$

In this problem, the decision variables and the objective function are identical to problem (P) from Chapter 2. A subscript m is included on the sparing factor s_m, the LQ dose-response parameter $\rho_m = \beta_m/\alpha_m$, the conventional number of treatment sessions T_m, the dose per session δ_m in this conventional plan, and the shorthand C_m for the tolerable BED of this conventional plan. These parameters appear in constraint (3.2) for the mth OAR. The constraint ensures that the BED of the proposed dosing plan is tolerable by each OAR. As in problem (P), we will first solve problem (Q) when the number of sessions $N \geq 1$ is fixed. We write that problem as QCQP

$$(\text{Q}(N))\ f^\star(N) = \max_{\vec{d}}\ f(N, \vec{d}) = \alpha_0 \sum_{t=1}^{N} d_t + \beta_0 \sum_{t=1}^{N} (d_t)^2 - \tau(N) \tag{3.5}$$

$$s_m \sum_{t=1}^{N} d_t + s_m^2 \rho_m \sum_{t=1}^{N} (d_t)^2 \leq T_m \delta_m (1 + \rho_m \delta_m),\ m \in \mathcal{M} \tag{3.6}$$

$$\vec{d} \geq 0. \tag{3.7}$$

Proposition 3.1 generalizes Proposition 2.1 to this problem.

As in Chapter 2, the condition in the first claim of this proposition is believed to be met for cancers such as head-and-neck and lung. The condition in the second claim is expected to hold for prostate cancer. The proof of the third claim is similar to the first two, except that some strict inequalities are replaced with their weak counterparts. So we only prove the first two. As in Chapter 2, we provide several different proofs as they include distinct insights.

Proposition 3.1 *We make three separate claims.*

Claim 1 *Suppose* $\frac{\alpha_0}{\beta_0} > \max_{m \in \mathcal{M}} \frac{1}{s_m} \frac{\alpha_m}{\beta_m}$. *Let*

$$d_m^{\star}(N) = \frac{-1 + \sqrt{1 + \frac{4\rho_m C_m}{N}}}{2 s_m \rho_m}, \ \forall m \in \mathcal{M}. \tag{3.8}$$

Then, the unique optimal solution to problem $(Q(N))$ *is given by* $d_1 = d_2 = \cdots = d_N = c^{\star}(N)$, *where*

$$c^{\star}(N) = \min_{m \in \mathcal{M}} d_m^{\star}(N). \tag{3.9}$$

We call this an equal-dosage plan.

Claim 2 *Suppose* $\frac{\alpha_0}{\beta_0} < \min_{m \in \mathcal{M}} \frac{1}{s_m} \frac{\alpha_m}{\beta_m}$. *Let*

$$d_m^{\circ} = \frac{-1 + \sqrt{1 + 4\rho_m C_m}}{2 s_m \rho_m}, \ \forall m \in \mathcal{M}, \tag{3.10}$$

and

$$\gamma^{\circ} = \min_{m \in \mathcal{M}} d_m^0. \tag{3.11}$$

Then, the only optimal solutions to problem $(Q(N))$ *are such that the dose in one session is* γ° *and the doses in the other* $N-1$ *sessions are* 0. *We call these single-dosage plans.*

Claim 3 *Suppose* $\frac{\alpha_0}{\beta_0} = \max_{m \in \mathcal{M}} \frac{1}{s_m} \frac{\alpha_m}{\beta_m}$. *Then, the equal-dosage plan described in Claim 1 is optimal (although not necessarily uniquely so). Suppose* $\frac{\alpha_0}{\beta_0} = \min_{m \in \mathcal{M}} \frac{1}{s_m} \frac{\alpha_m}{\beta_m}$. *Then, the single-dosage plans described in Claim 2 are optimal (although not necessarily uniquely so).*

3.1 Optimal Solution Using Cauchy–Schwarz

As in problem (P(N)), constraint (3.6) must be active for at least one OAR, at an optimal solution to problem (Q(N)). Suppose the constraint is active for OAR m. Then, $\sum_{t=1}^{N}(d_t)^2 = \frac{C_m - s_m \sum_{t=1}^{N} d_t}{s_m^2 \rho_m}$. Similar to (2.11), the objective function of (Q(N)) can be expressed as

$$\left(\sum_{t=1}^{N} d_t\right) \beta_0 \left(\frac{\alpha_0}{\beta_0} - \frac{\alpha_m}{s_m \beta_m}\right) + \frac{\beta_0 C_m}{s_m^2 \rho_m} - \tau(N). \tag{3.12}$$

Thus, for the reason explained in Section 2.1, an equal-dosage solution maximizes this objective, since $\frac{\alpha_0}{\beta_0} > \max_{m \in \mathcal{M}} \frac{1}{s_m} \frac{\alpha_m}{\beta_m} \geq \frac{\alpha_m}{s_m \beta_m}$. In fact, more strongly, any dosing plan that is not equal dosage and yet renders constraint (3.6) active for OAR m has a strictly smaller objective value than that of this equal-dosage solution. This equal-dosage solution is given by $d_m^\star(N)$ as in (3.8). Further, the left-hand side of constraint (3.6) is increasing in the dose administered in any session. Thus, any equal-dosage solution that administers a dose larger than $d_m^\star(N)$ per session is infeasible. Repeating this argument for each OAR in $\mathcal{M}$ one by one, we see that the equal-dosage solution that administers dose $c^\star(N)$ as in (3.9) in each session is optimal. This proves Claim 1.

Similarly, for the reason explained in Section 2.1, a single-dosage solution maximizes this objective, since $\frac{\alpha_0}{\beta_0} < \min_{m \in \mathcal{M}} \frac{1}{s_m} \frac{\alpha_m}{\beta_m} \leq \frac{\alpha_m}{s_m \beta_m}$. In fact, more strongly, any dosing plan that is not single dosage and yet renders constraint (3.6) active for OAR m has a strictly smaller objective value than that of this single-dosage solution. This single-dosage solution is given by d_m° as in (3.10). Further, the left-hand side of constraint (3.6) is increasing in the dose administered in any session. Thus, any single-dosage solution that administers a dose larger than d_m° is infeasible. Repeating this argument for each OAR in $\mathcal{M}$ one by one, we see that the single-dosage solution that administers dose γ° as in (3.11) in a single session is optimal. This proves Claim 2.

3.2 Algebraic Derivation

For Claim 1, as in the single OAR case, consider an unequal-dosage plan $\vec{d} = (d_1, \ldots, d_N)$, with $d_j \neq d_k$ for two distinct sessions j, k, that is feasible to (Q(N)). Construct an alternative equal-dosage plan where the dose in each session is the average $b = \sum_{t=1}^{N} d_t / N$. By algebra identical to the single OAR case, this plan is strictly feasible to each OAR $m \in \mathcal{M}$. Let $\epsilon_m > 0$, for each

$m \in \mathcal{M}$, be such that the equal-dosage plan $(b+\epsilon_m, \ldots, b+\epsilon_m)$ renders (3.2) active for OAR m. Let $\epsilon^\star = \min_{m \in \mathcal{M}} \epsilon_m$ and $m^\star \in \operatorname{argmin}_{m \in \mathcal{M}} \epsilon_m$. Then, the equal-dosage plan $\vec{w} = (b+\epsilon^\star, \ldots, b+\epsilon^\star)$ is feasible and also renders constraint (3.2) active for OAR $m^\star$. Thus, $s_{m^\star} N(b+\epsilon^*) + (s_{m^\star})^2 \rho_{m^\star} N(b+\epsilon^*)^2 = C_{m^\star}$. Let $\Delta(m^\star)$ denote $\alpha_0 - \frac{\beta_0}{s_{m^\star}\rho_{m^\star}}$ for brevity. Using this, the objective function in problem (Q(N)) can be expressed, with algebra identical to (3.12), as

$$\begin{aligned}
f(N,\vec{w}) &= \Delta(m^\star)\sum_{t=1}^{N} d_t + N\epsilon^\star\Delta(m^\star) + \frac{C_{m^\star}\beta_0}{(s_{m^\star})^2\rho_{m^\star}} - \tau(N) \\
&\geq \Delta(m^\star)\sum_{t=1}^{N} d_t + N\epsilon^\star\Delta(m^\star) \\
&\quad + \frac{\left(s_{m^\star}\sum_{t=1}^{N} d_t + (s_{m^\star})^2\rho_{m^\star}\sum_{t=1}^{N}(d_t)^2\right)\beta_0}{(s_{m^\star})^2\rho_{m^\star}} - \tau(N) \\
&= \alpha_0\sum_{t=1}^{N} d_t + \beta_0\sum_{t=1}^{N}(d_t)^2 - \tau(N) + N\epsilon^\star\Delta(m^\star) \\
&= f(N,\vec{d}) + N\epsilon^\star\Delta(m^\star) > f(N,\vec{d}).
\end{aligned}$$

Here, the first inequality holds because $s_{m^\star}\sum_{t=1}^{N} d_t + (s_{m^\star})^2\rho_{m^\star}\sum_{t=1}^{N}(d_t)^2 \leq C_{m^\star}$ as $\vec{d}$ is feasible to (3.6). The last strict inequality above holds because $\epsilon^\star > 0$ by construction and $\Delta(m^\star) > 0$ by assumption. This strict inequality implies that no unequal-dosage plan can be optimal. Claim 1 then follows because formula (3.9) provides the best equal-dosage solution as explained in Section 3.1.

We prove Claim 2 by contradiction. As in the single OAR case, suppose dosing plan $\vec{d} = (d_1, \ldots, d_N) \geq \vec{0}$ that includes at least two sessions with positive doses is optimal. By algebra identical to the single OAR case, the alternative dosing plan $\left(\sqrt{\sum_{t\in\mathcal{T}}(d_t)^2}, \underbrace{0,\ldots,0}_{N-1 \text{ times}}\right)$ is strictly feasible. Thus, there is an $\epsilon_m > 0$ such that $\vec{w}^m = \left(\sqrt{\sum_{t\in\mathcal{T}}(d_t)^2 + \epsilon_m}, \underbrace{0,\ldots,0}_{N-1 \text{ times}}\right)$ renders constraint (3.6) active, for OAR $m \in \mathcal{M}$. Let $\epsilon^\star = \min_{m\in\mathcal{M}} \epsilon_m$ and $m^\star \in \operatorname{argmin}_{m\in\mathcal{M}} \epsilon_m$ as in the above proof of Claim 1. Then, the single-dosage plan $\vec{w} = \left(\sqrt{\sum_{t\in\mathcal{T}}(d_t)^2 + \epsilon^\star}, \underbrace{0,\ldots,0}_{N-1 \text{ times}}\right)$ is feasible to problem (Q(N))

and renders constraint (3.6) active, for OAR $m^\star$. Let $\zeta > 0$ be such that $\sqrt{\sum_{t\in\mathcal{T}}(d_t)^2+\zeta} = \sum_{t\in\mathcal{T}} d_t$. Suppose constraint (3.6) is active at $\vec{d}$, for some OAR $n^\star \in \mathcal{M}$. Such an OAR exists because $\vec{d}$ is optimal. Then,

$$
\begin{aligned}
s_{n^\star}\sum_{t\in\mathcal{T}} d_t + (s_{n^\star})^2\rho_{n^\star}\sum_{t\in\mathcal{T}}(d_t)^2 &= C_{n^\star}\\
&\geq s_{n^\star}\left(\sqrt{\sum_{t\in\mathcal{T}}(d_t)^2}+\epsilon^\star\right)\\
&\quad + (s_{n^\star})^2\rho_{n^\star}\left(\sum_{t\in T}(d_t)^2+(\epsilon^\star)^2+2\epsilon^\star\sqrt{\sum_{t\in\mathcal{T}}(d_t)^2}\right). \qquad \text{(feasibility of } \vec{w}\text{)}
\end{aligned}
$$

Then, algebra identical to the single OAR case from Section 2.5 implies that $\zeta > \epsilon^\star$.

Since $\vec{d}$ satisfies constraint (3.6) for OAR $m^\star$, we know that $\sum_{t=1}^{N}(d_t)^2 \leq \frac{C_{m^\star}-s_{m^\star}\sum_{t=1}^{N} d_t}{(s_{m^\star})^2\rho_{m^\star}}$. Using this, the objective value of $\vec{d}$ is

$$
\begin{aligned}
f(N,\vec{d}) &= \alpha_0\sum_{t=1}^{N} d_t + \beta_0\sum_{t=1}^{N}(d_t)^2 - \tau(N)\\
&\leq \left(\alpha_0 - \frac{\beta_0}{s_{m^\star}\rho_{m^\star}}\right)\left(\sum_{t=1}^{N} d_t\right) + \frac{C_{m^\star}\beta_0}{s_{m^\star}^2\rho_{m^\star}} - \tau(N)\\
&\qquad \text{(algebra similar to (3.12))}\\
&= \left(\alpha_0 - \frac{\beta_0}{s_{m^\star}\rho_{m^\star}}\right)\left(\sum_{t\in\mathcal{T}} d_t\right) + \frac{C_{m^\star}\beta_0}{(s_{m^\star})^2\rho_{m^\star}} - \tau(N)\\
&\qquad \text{(since } d_t = 0 \text{ when } t \notin \mathcal{T}\text{)}\\
&= \left(\alpha_0 - \frac{\beta_0}{s_{m^\star}\rho_{m^\star}}\right)\left(\sqrt{\sum_{t\in\mathcal{T}}(d_t)^2+\zeta}\right) + \frac{C_{m^\star}\beta_0}{(s_{m^\star})^2\rho_{m^\star}} - \tau(N)\\
&\qquad \left(\text{from } \sum_{t\in\mathcal{T}} d_t = \sqrt{\sum_{t\in\mathcal{T}}(d_t)^2+\zeta}\right)\\
&< \left(\alpha_0 - \frac{\beta_0}{s_{m^\star}\rho_{m^\star}}\right)\left(\sqrt{\sum_{t\in\mathcal{T}}(d_t)^2}+\epsilon^\star\right) + \frac{C_{m^\star}\beta_0}{(s_{m^\star})^2\rho_{m^\star}} - \tau(N)\\
&\qquad \left(\epsilon^\star < \zeta \text{ and } \alpha_0 - \frac{\beta_0}{s_{m^\star}\rho_{m^\star}} < 0\right)\\
&= f(N,\vec{w}). \qquad \text{(algebra similar to (3.12))}
\end{aligned}
$$

This contradicts the optimality of the two-or-more dosing plan $\vec{d}$. Since $\vec{d}$ was an arbitrary two-or-more dosing plan, this demonstrates that no such plan can be optimal. Since at least one dose must be positive at optimality, Claim 2 follows because formula (3.11) provides the best single-dosage solution as explained in Section 3.1.

3.3 Optimal Solution Using Karush–Kuhn–Tucker Conditions

The Karush–Kuhn–Tucker (KKT) necessary conditions for optimality in problem (Q(N)) are given by

$$-\begin{bmatrix} \alpha_0 + 2\beta_0 d_1 \\ \vdots \\ \alpha_0 + 2\beta_0 d_N \end{bmatrix} + \sum_{m \in \mathcal{M}} \lambda_m s_m \begin{bmatrix} 1 + 2 s_m \rho_m d_1 \\ \vdots \\ 1 + 2 s_m \rho_m d_N \end{bmatrix} - \sum_{t=1}^{N} \mu_t \vec{e}_t = \vec{0} \tag{3.13}$$

$$(3.6) - (3.7) \tag{3.14}$$

$$\left(C_m - s_m \sum_{t=1}^{N} d_t - (s_m)^2 \rho_m \sum_{t=1}^{N} (d_t)^2 \right) \lambda_m = 0 \tag{3.15}$$

$$\mu_t d_t = 0,\ t = 1\colon N \tag{3.16}$$

$$\lambda_m \geq 0,\ m \in \mathcal{M} \tag{3.17}$$

$$\mu_t \geq 0,\ t = 1\colon N. \tag{3.18}$$

Here, λ_m is the Lagrange multiplier associated with constraint (3.6) and the rest of the notation is the same as Section 2.3. Recall from that section that, for each $k = 1\colon N$, $\vec{d}(k)$ denotes $(d_1, d_2, \ldots, d_k, 0, 0, \ldots, 0)$, with $d_1 > 0, d_2 > 0, \ldots, d_k > 0$.

KKT condition (3.16) implies that $\mu_1 = \cdots = \mu_k = 0$. Thus, (3.13) yields

$$-\alpha_0 - 2\beta_0 d_t + \sum_{m \in \mathcal{M}} \lambda_m s_m (1 + 2 s_m \rho_m d_t) = 0,\ t = 1\colon k \tag{3.19}$$

$$-\alpha_0 + \sum_{m \in \mathcal{M}} \lambda_m s_m = \mu_t,\ t = k+1\colon N. \tag{3.20}$$

We rewrite (3.19) as

$$2d_t \left(-\beta_0 + \sum_{m \in \mathcal{M}} \lambda_m (s_m)^2 \rho_m \right) = \alpha_0 - \sum_{m \in \mathcal{M}} \lambda_m s_m,\ t = 1\colon k. \tag{3.21}$$

Since the right-hand side of (3.21) does not depend on t, we obtain

$$d_1\left(-\beta_0+\sum_{m\in\mathcal{M}}\lambda_m(s_m)^2\rho_m\right)=\cdots=d_k\left(-\beta_0+\sum_{m\in\mathcal{M}}\lambda_m(s_m)^2\rho_m\right). \quad (3.22)$$

We prove by contradiction that $-\beta_0+\sum_{m\in\mathcal{M}}\lambda_m(s_m)^2\rho_m\neq 0$. This would imply from (3.22) that $d_1=\cdots=d_k$. So suppose $\beta_0=\sum_{m\in\mathcal{M}}\lambda_m(s_m)^2\rho_m$. Substituting this into (3.21) yields $\alpha_0=\sum_{m\in\mathcal{M}}\lambda_m s_m$. Thus,

$$\begin{aligned}\frac{\alpha_0}{\beta_0}&=\frac{\sum_{m\in\mathcal{M}}\lambda_m s_m}{\sum_{m\in\mathcal{M}}\lambda_m(s_m)^2\rho_m}=\frac{\sum_{m\in\mathcal{M}}\lambda_m(s_m)^2\rho_m\left(\frac{1}{s_m}\frac{\alpha_m}{\beta_m}\right)}{\sum_{m\in\mathcal{M}}\lambda_m(s_m)^2\rho_m}\\&=\sum_{m\in\mathcal{M}}a_m\left(\frac{1}{s_m}\frac{\alpha_m}{\beta_m}\right).\end{aligned} \quad (3.23)$$

Here, we have defined the notation $a_m=\frac{\lambda_m(s_m)^2\rho_m}{\sum_{m\in\mathcal{M}}\lambda_m(s_m)^2\rho_m}$, and thus $a_m\geq 0$ with $\sum_{m\in\mathcal{M}}a_m=1$. Since $\lambda_m\geq 0$ and not all of these Lagrange multipliers can be 0 (this can be established from (3.19) and (3.20)), (3.23) implies that $\frac{\alpha_0}{\beta_0}$ is a convex combination of $\frac{1}{s_m}\frac{\alpha_m}{\beta_m}$. But this contradicts the supposition in the first two claims of the proposition that either $(\alpha_0/\beta_0)>\alpha_m/(s_m\beta_m)$, for all $m\in\mathcal{M}$, or $(\alpha_0/\beta_0)<\alpha_m/(s_m\beta_m)$, for all $m\in\mathcal{M}$. This shows that $d_1=\cdots=d_k$, which we denote by $d^\star(k)>0$.

Equation (3.19) now implies that

$$\sum_{m\in\mathcal{M}}\lambda_m s_m=\alpha_0+2\beta_0 d^\star(k)-2d^\star(k)\sum_{m\in\mathcal{M}}\lambda_m(s_m)^2\rho_m, \quad (3.24)$$

and substituting this into (3.20) yields

$$\mu_t=2d^\star(k)\left(\beta_0-\sum_{m\in\mathcal{M}}\lambda_m(s_m)^2\rho_m\right),\ t=k+1:N. \quad (3.25)$$

In fact, since the right-hand side in (3.25) does not depend on t, we observe that $\mu_{k+1}=\cdots=\mu_N$ and denote this value by the real number μ. Now we consider the specific implications of the relative values of α_0/β_0 and $\alpha_m/(s_m\beta_m)$, thereby leading to separate proofs of the first two claims in the proposition.

For the first claim, we have $\alpha_0/\beta_0>\alpha_m/(s_m\beta_m)$, for all $m\in\mathcal{M}$. Suppose $k<N$. We claim that $\mu\geq 0$ implies $d^\star(k)<0$. To see this, first observe from (3.20) that

$$\sum_{m\in\mathcal{M}}\lambda_m s_m\geq\alpha_0. \quad (3.26)$$

Without loss of generality, suppose that $\min_{m \in \mathcal{M}} s_m \rho_m = s_1 \rho_1$. Thus,

$$\beta_0 - \sum_{m \in \mathcal{M}} \lambda_m (s_m)^2 \rho_m \leq \beta_0 - s_1 \rho_1 \sum_{m \in \mathcal{M}} \lambda_m s_m \leq \beta_0 - s_1 \rho_1 \alpha_0 < 0. \quad (3.27)$$

Here, the first inequality holds because $s_1 \rho_1$ is the smallest among all $s_m \rho_m$ values. The second inequality follows from (3.26). The strict inequality holds because $\frac{\alpha_0}{\beta_0} > \max_{m \in \mathcal{M}} \frac{1}{s_m \rho_m} = \frac{1}{\min_{m \in \mathcal{M}} s_m \rho_m} = \frac{1}{s_1 \rho_1}$. Substituting this in (3.25) and the supposition that $\mu \geq 0$ imply that $d^\star(k) < 0$, as claimed. Consequently, for $k < N$, there are no solutions that satisfy the KKT condition (3.18). Thus, a KKT solution must come from the case $k = N$. In other words, the only possible KKT solutions are of the form $(d^\star(N), d^\star(N), \ldots, d^\star(N))$, that is, equal-dosage. Consequently, an equal-dosage solution with the largest objective function value must be optimal. But recall from the previous two sections that this best equal-dosage solution is given by formula (3.9). This proves the first claim in the proposition.

For the second claim, we have $\alpha_0/\beta_0 < \alpha_m/(s_m \beta_m)$, for all $m \in \mathcal{M}$. Recall that there are N different solutions that could satisfy all KKT conditions. These candidate KKT solutions are of the form

$$(\underbrace{d^\star(k), \ldots, d^\star(k)}_{k \text{ times}}, 0, \ldots, 0).$$

An argument similar to the previous two sections demonstrates that the best value of $d^\star(k)$ is given by

$$d^\star(k) = \min_{m \in \mathcal{M}} \left(\frac{-1 + \sqrt{1 + \frac{4 \rho_m C_m}{k}}}{2 s_m \rho_m} \right), \quad (3.28)$$

for $k = 1 : N$. In order to find the best among these N solutions, we compare their objective function values in problem (Q(N)). The objective value of the kth solution equals $\alpha_0 k d^\star(k) + \beta_0 k (d^\star(k))^2 - \tau(N)$. Since the proliferation term $\tau(N)$ does not depend on k, it suffices to focus on the part $\alpha_0 k d^\star(k) + \beta_0 k (d^\star(k))^2$ without proliferation. This expression is strictly decreasing in real numbers $k > 0$, since $\alpha_0/\beta_0 < \alpha_m/(s_m \beta_m)$, for all $m \in \mathcal{M}$. We omit the proof here, because a similar calculus-based argument was presented in Section 2.6.1. This implies that, among the N candidate KKT solutions described above, the one where $k = 1$ produces the largest objective value. It must therefore be optimal as long as it satisfies all KKT conditions with appropriate values of Lagrange multipliers. Thus, it remains to find appropriate Lagrange multipliers. Toward this end, without loss of generality, suppose that OAR

$m = 1$ attains the minimum over $\mathcal{M}$ in (3.28) when $k = 1$. Thus, administering dose $d^\star(1)$ in a single session renders constraint (3.6) active for OAR $m = 1$. Now consider the Lagrange multipliers

$$\lambda_1 = \frac{\alpha_0 + 2\beta_0 d^\star(1)}{s_1(1 + 2s_1\rho_1 d^\star(1))}; \ \lambda_m = 0, \ m \in \mathcal{M} \setminus \{1\}$$

$$\mu_1 = 0; \ \mu_t = \frac{2d^\star(1)(\beta_0 - \alpha_0 s_1 \rho_1)}{1 + 2s_1\rho_1 d^\star(1)}, \ t = 2 \colon N.$$

These expressions were motivated by the Lagrange multipliers from the single OAR case in Section 2.3. It is easy to check, just as in Section 2.3, that these satisfy the KKT conditions (3.13)–(3.18). Finally, observe that the solution (3.28) with $k = 1$ precisely matches formula (3.11) in the proposition. This proves the second claim in the proposition.

3.4 Conversion into a Two-Variable Problem

We first state a generalization of Lemma 2.3. Its proof is identical to that of Lemma 2.3 and hence is omitted.

Lemma 3.2 *The N-variable problem $(Q(N))$ is equivalent to the two-variable problem*

$$(2VAR(N)) \ f^\star(N) = \max_{x,y} \ \alpha_0 x + \beta_0 y - \tau(N) \tag{3.29}$$

$$s_m x + s_m^2 \rho_m y \leq C_m, \ m \in \mathcal{M} \tag{3.30}$$

$$\sqrt{y} \leq x \tag{3.31}$$

$$x \leq \sqrt{Ny} \tag{3.32}$$

$$x \geq 0, y \geq 0 \tag{3.33}$$

in the sense summarized in three claims below.

Claim 1 *If $\vec{d}$ is feasible to $(Q(N))$, then the solution $x = \sum_{t=1}^N d_t$ and $y = \sum_{t=1}^N d_t^2$ is feasible to problem $(2VAR(N))$; solutions $\vec{d}$ and (x, y) have identical objective function values in the two problems.*

Claim 2 *Conversely, if (x, y) is feasible to $(2VAR(N))$, Lemma 2.2 provides a method to construct a feasible solution $\vec{d}$ to problem $(Q(N))$ such that the two solutions have identical objective function values in the two problems.*

Claim 3 *An optimal solution to one problem corresponds to an optimal solution to the other problem, and the two optimal objective values are equal.*

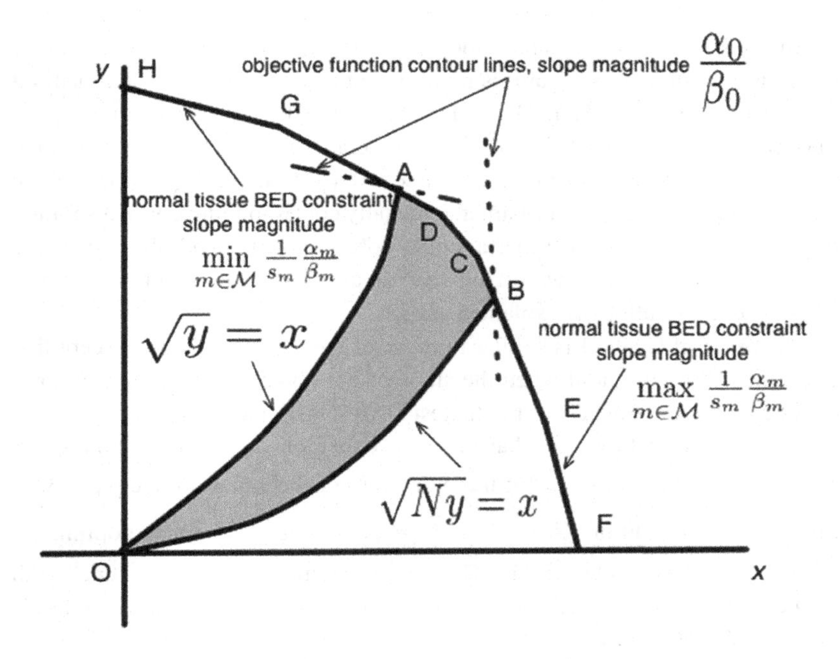

Figure 3.1 Feasible region (shaded in gray) and objective function contour lines for problem (2VAR(N)). Reused with permission from [61, figure 6].

We will not utilize problem (2VAR(N)) as a computational device, for now. Instead, we now prove Proposition 3.1 with the help of Figure 3.1, which illustrates the feasible region of problem (2VAR(N)). The proof is similar to that in the single OAR case from Section 2.3.

As in Figure 2.1 for the single OAR case from Section 2.3, we want to push the objective function contour lines outward away from the origin in the positive $x - y$ quadrant, as much as possible. Thus, optimal solutions can only occur at the boundary formed by segments AD, DC, and CB.

If $\frac{\alpha_0}{\beta_0} > \max_{m \in \mathcal{M}} \frac{1}{s_m} \frac{\alpha_m}{\beta_m}$, then the slope of the objective function contour line (dotted in the figure) is steeper than the slopes of all OAR constraint boundaries (segments HG, GA, AD, DC, CB, BE, and EF in the figure). Specifically, it is steeper than the slope of segment CE. The unique optimal solution then occurs at point B, where $x > 0, y > 0$ and $x = \sqrt{Ny}$. Thus, from Claim 2 in Lemma 2.2 from Chapter 2, the corresponding optimal solution to problem (Q(N)) must have $d_1 = d_2 = \cdots = d_N$. Claim 1 then follows because $c^{\star}(N)$ given by formula (3.9) is the best equal-dosage solution, as explained earlier.

Similarly, if $\frac{\alpha_0}{\beta_0} < \frac{1}{s}\frac{\alpha}{\beta}$, then the slope of the objective function contour lines (dash-dotted in the figure) is flatter than the slopes of all OAR constraint boundaries (segments HG, GA, AD, DC, CB, BE, and EF in the figure). Specifically, it is flatter than the slope of segment GA. The unique optimal solution then occurs at point A, where $x > 0, y > 0$ and $x = \sqrt{y}$. Thus, from Claim 1 in Lemma 2.2 from Chapter 2, the only corresponding optimal solution to problem (Q(N)) is such that a single dose is positive and all other doses are 0. Claim 2 then follows because γ° given by formula (3.11) is the best single-dosage solution, as explained earlier.

The proof of Claim 3 is similar to those of the first two claims, except that uniqueness does not hold when the slope of the objective function contour is either equal to the steepest or the flattest OAR constraint boundary.

As an aside, it turns out that unlike Proposition 2.1 the condition $\frac{\alpha_0}{\beta_0} \geq \max_{m \in \mathcal{M}} \frac{1}{s_m}\frac{\alpha_m}{\beta_m}$ is not necessary for the optimality of the equal-dosage plan (3.9). Similarly, the condition $\frac{\alpha_0}{\beta_0} \leq \min_{m \in \mathcal{M}} \frac{1}{s_m}\frac{\alpha_m}{\beta_m}$ is not necessary for the optimality of the single-dosage plan (3.11). These two statements can be explained with the help of Figure 3.1. Specifically, the figure shows that the slope magnitude of the dash-dotted objective function contour line does not have to be flatter than the flattest OAR constraint boundary HG for the single-dosage solution corresponding to point A to be optimal. It would be sufficient to be flatter than the second flattest OAR constraint boundary GD. Similarly, the slope magnitude of the dotted objective function contour line does not have to be steeper than the steepest OAR constraint boundary EF for the equal-dosage solution corresponding to point B to be optimal. It would be sufficient to be steeper than the second steepest OAR constraint boundary CE.

3.5 Solution of an Edge Case

Proposition 3.1 leaves open the case where

$$\min_{m \in \mathcal{M}} \frac{1}{s_m}\frac{\alpha_m}{\beta_m} < \frac{\alpha_0}{\beta_0} < \max_{m \in \mathcal{M}} \frac{1}{s_m}\frac{\alpha_m}{\beta_m}. \tag{3.34}$$

This situation did not arise in Chapter 2. When (3.34) occurs, points C or D can be optimal in Figure 3.1. These solutions are neither single dosage nor equal dosage because they do not lie on the curve $x = \sqrt{y}$ or on the curve $x = \sqrt{Ny}$. We present an example with $N = 2$ to illustrate this scenario.

Example 3.3 Consider the following parameters: $\alpha_0 = 1$ Gy^{-1}; $\beta_0 = 0.2$ Gy^{-2}; $s_1 = s_2 = 1$; $\rho_1 = 1/6 = 0.1666$ Gy^{-1}; $\rho_2 = 0.3571$ Gy^{-1};

$T_1 = T_2 = 35$; $\delta_1 = 1.0857$ Gy; $\delta_2 = 1.4857$ Gy. For these parameters, we have

$$\min\left\{\frac{1}{s_1}\frac{\alpha_1}{\beta_1}, \frac{1}{s_2}\frac{\alpha_2}{\beta_2}\right\} = \min\left\{\frac{1}{\rho_1 s_1}, \frac{1}{\rho_2 s_2}\right\} = \min\{6, 2.8003\}$$
$$= 2.8003 < \frac{\alpha_0}{\beta_0} = \frac{1}{0.2} = 5 \text{ Gy} < \max\left\{\frac{1}{s_1}\frac{\alpha_1}{\beta_1}, \frac{1}{s_2}\frac{\alpha_2}{\beta_2}\right\} = 6 \text{ Gy}.$$

Thus, condition (3.34) holds for these parameters. Substituting these parameters into (Q(N)) yields

$$\max\ d_1 + 0.2d_1^2 + d_2 + 0.2d_2^2 \tag{3.35}$$
$$d_1 + 0.1666d_1^2 + d_2 + 0.1666d_2^2 \leq 44.8762 \tag{3.36}$$
$$d_1 + 0.3571d_1^2 + d_2 + 0.3571d_2^2 \leq 79.5885 \tag{3.37}$$
$$d_1 \geq 0,\ d_2 \geq 0. \tag{3.38}$$

The constant term $\tau(2)$ was omitted from the objective function of this problem because it has no effect on the choice of optimal d_1, d_2 values. This maximization problem has only two optimal solutions: $d_1^\star \approx 13.45$ Gy, $d_2^\star \approx 1.05$ Gy and $d_1^\star \approx 1.05$ Gy, $d_2^\star \approx 13.45$ Gy (we will explain a simple method below to derive such solutions). Neither of these two solutions is single dosage or equal dosage.

Lemma 3.4 demonstrates how one can solve (Q(N)) when (3.34) occurs. In fact, the solution method actually works regardless of the relative values of $\frac{\alpha_0}{\beta_0}$, $\min\limits_{m\in\mathcal{M}} \frac{\alpha_m}{s_m\beta_m}$, and $\max\limits_{m\in\mathcal{M}} \frac{\alpha_m}{s_m\beta_m}$, although it is not needed when (3.34) does not hold (by Proposition 3.1). The method relies on the equivalent reformulation (2VAR(N)) of (Q(N)) as established in Lemma 3.2. But first recall that although (2VAR(N)) includes only two decision variables and a linear objective, constraints $\sqrt{y} \leq x$ and $x \leq \sqrt{Ny}$ are nonlinear. Moreover, the constraint $\sqrt{y} \leq x$ renders (2VAR(N)) nonconvex. The lemma below circumvents this difficulty.

Lemma 3.4 *The pair $(x^\star(N), y^\star(N))$ is optimal to (2VAR(N)) if, and only if, it is optimal to the two-variable LP*

$$\text{(2VARLP(}N\text{))} \quad f^\star(N) = \max_{x,y}\ \alpha_0 x + \beta_0 y - \tau(N)$$
$$s_m x + s_m^2 \rho_m y \leq C_m,\ m \in \mathcal{M}$$
$$c^\star(N)x \leq y \tag{3.39}$$

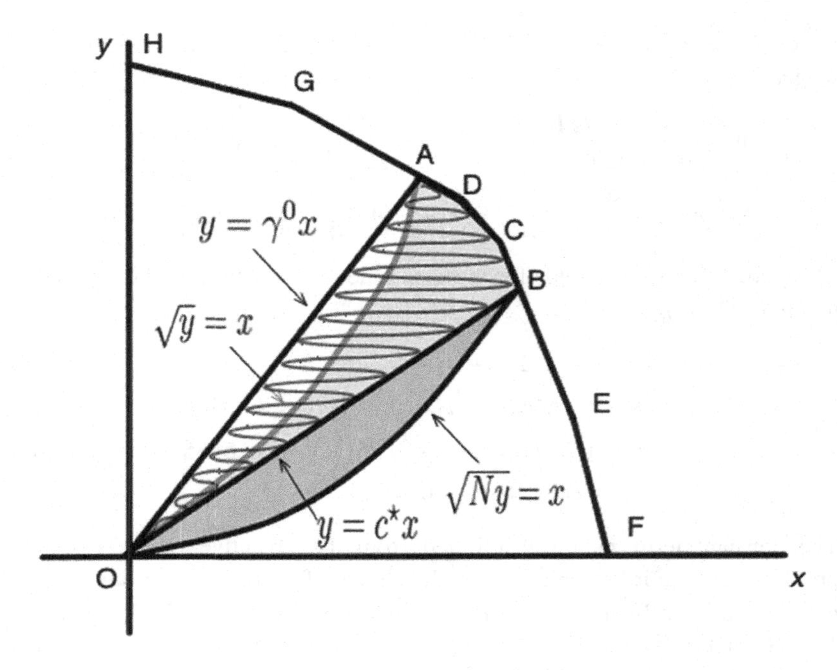

Figure 3.2 (2VAR(N)) can be equivalently reformulated as (2VARLP(N)) because the shaded gray horn-shaped feasible region of the former can be replaced with the hatched cone-shaped region of the latter. Adapted from [102] per reuse policy available at www.elsevier.com/about/policies/copyright.

$$y \leq \gamma^{\circ} x \tag{3.40}$$

$$x \geq 0,\ y \geq 0.$$

Proof Refer to Figure 3.2. The gray horn-shaped shaded area depicts the feasible region of (2VAR(N)). Optimal solutions can only occur at the boundary ADCB. Problem (2VARLP(N)) replaces the feasible region of (2VAR(N)) with a cone-shaped hatched region. Edge OB corresponds to the boundary $y = c^{\star}(N)x$ of constraint (3.39) and includes the best equal-dosage solution B. Edge OA of this cone-shaped region corresponds to the boundary $y = \gamma^{\circ}x$ of constraint (3.40) and includes the best single-dosage solution A. As such, replacing the horn-shaped feasible region with the cone-shaped feasible region does not exclude any (x, y) pairs that could have been optimal to (2VAR(N)) and does not include any new (x, y) pairs with strictly larger objective function values than any feasible solution to (2VAR(N)). This completes the proof. □

3.6 Optimal Number of Sessions

Proposition 3.1 establishes that it is optimal to administer dose (3.11) in a single session and not to administer any dose in other sessions, when $\frac{\alpha_0}{\beta_0} \leq \min_{m \in \mathcal{M}} \frac{1}{s_m} \frac{\alpha_m}{\beta_m}$. This holds in problem (Q(N)) regardless of N. Thus, the optimal number of sessions in problem (Q) equals 1 with a dose given by (3.11). Our next result thus focuses on the case $\frac{\alpha_0}{\beta_0} \geq \max_{m \in \mathcal{M}} \frac{1}{s_m} \frac{\alpha_m}{\beta_m}$.

Theorem 3.5 *Suppose $\frac{\alpha_0}{\beta_0} \geq \max_{m \in \mathcal{M}} \frac{1}{s_m} \frac{\alpha_m}{\beta_m}$. Then, there exists an integer $\hat{N} \geq 2$ such that $f^\star(\hat{N}) < f^\star(\hat{N} - 1)$. That is, the objective function in* (3.41) *eventually decreases. Let $N^\star$ denote the smallest such $\hat{N}$. It is optimal to administer $N_{optimal} = \min\{N^\star - 1, N_{max}\}$ sessions with dose $c^\star(N_{optimal})$ each.*

We will prove this theorem in the next subsection. Observe, however, that the discussion in this section thus far has left out the case where $\min_{m \in \mathcal{M}} \frac{1}{s_m} \frac{\alpha_m}{\beta_m} < \frac{\alpha_0}{\beta_0} < \max_{m \in \mathcal{M}} \frac{1}{s_m} \frac{\alpha_m}{\beta_m}$. The structure of $f^\star(N)$ is not known in that case. As such, it is not known whether or not a counterpart of Theorem 3.5 holds in that case. Nevertheless, recall that problem (Q(N)) can be solved exactly and thus $f^\star(N)$ can be calculated via the two-variable LP described in Lemma 3.4 from Section 3.5, even when $\min_{m \in \mathcal{M}} \frac{1}{s_m} \frac{\alpha_m}{\beta_m} < \frac{\alpha_0}{\beta_0} < \max_{m \in \mathcal{M}} \frac{1}{s_m} \frac{\alpha_m}{\beta_m}$. An optimal number of sessions can then be determined by finding an integer $1 \leq N \leq N_{\max}$ that maximizes the value of $f^\star(N)$ calculated that way. This discussion yields the procedure outlined in Algorithm 1 for solving problem (Q).

3.6.1 Solution Using Calculus

We substitute the optimal dose $c^\star(N)$ per session from (3.9) into the objective function of problem (Q) to obtain

$$
\begin{aligned}
f^\star = & \max_{\substack{1 \leq N \leq N_{\max} \\ N \text{ integer}}} f^\star(N) = \alpha_0 N c^\star(N) + \beta_0 N (c^\star(N))^2 - \tau(N) && (3.41) \\
= & \max_{\substack{1 \leq N \leq N_{\max} \\ N \text{ integer}}} \alpha_0 N \left(\min_{m \in \mathcal{M}} d_m^\star(N) \right) + \beta_0 N \left(\min_{m \in \mathcal{M}} d_m^\star(N) \right)^2 - \tau(N) \\
= & \max_{\substack{1 \leq N \leq N_{\max} \\ N \text{ integer}}} \left\{ \min_{m \in \mathcal{M}} \underbrace{\left(\alpha_0 N d_m^\star(N) + \beta_0 N (d_m^\star(N))^2 \right)}_{z_m(N)} - \tau(N) \right\} && (3.42) \\
= & \max_{\substack{1 \leq N \leq N_{\max} \\ N \text{ integer}}} z(N) - \tau(N). && (3.43)
\end{aligned}
$$

Algorithm 1 Exact solution of (Q)

1: Input - $\alpha_0, \beta_0, T_{\text{double}}$; set $\mathcal{M}$ of OAR; $T_m, \delta_m, \rho_m, s_m, \forall m \in \mathcal{M}$; $N_{\max}$.
2: Calculate $C_m = T_m\delta_m(1 + \rho_m\delta_m)$, for all $m \in \mathcal{M}$.
3: **if** $\frac{\alpha_0}{\beta_0} \le \min_{m\in\mathcal{M}} \frac{1}{s_m}\frac{\alpha_m}{\beta_m}$ **then** ▷ Single-dosage optimal (Proposition 3.1).
4: Output - dose γ° from (3.11) in a single session.
5: **else if** $\frac{\alpha_0}{\beta_0} \ge \max_{m\in\mathcal{M}} \frac{1}{s_m}\frac{\alpha_m}{\beta_m}$ **then** ▷ Equal-dosage optimal (Prop. 3.1).
6: best $= -\infty$.
7: $N_{\text{optimal}} = N_{\max}$. ▷ In case $f^\star$ never decreases in the for loop below.
8: **for** $N = 1 : N_{\max}$ **do**
9: Calculate $c^\star(N)$ from (3.9).
10: Calculate $f^\star(N) = \alpha_0 N c^\star(N) + \beta_0 N (c^\star(N))^2 - \frac{(N-1)\ln(2)}{T_{\text{double}}}$.
11: **if** $f^\star(N) \ge$ best **then**
12: best $= f^\star(N)$.
13: **else** ▷ This implies that $f^\star(N) < f^\star(N-1)$.
14: $N_{\text{optimal}} = N - 1$. ▷ Theorem 3.5.
15: Break out of for loop over N.
16: **end if**
17: **end for**
18: Output - dose $c^\star(N_{\text{optimal}})$ from (3.9) in each one of N_{optimal} sessions.
19: **else** ▷ Sufficient conditions in Proposition 3.1 not met.
20: best $= -\infty$.
21: **for** $N = 1 : N_{\max}$ **do**
22: Calculate $c^\star(N)$ from (3.9) and γ° from (3.11).
23: Solve (2VARLP(N)) to obtain optimal pair $(x^\star(N), y^\star(N))$.
24: Calculate $f^\star(N) = \alpha_0 x^\star(N) + \beta_0 y^\star(N) - \frac{(N-1)\ln(2)}{T_{\text{double}}}$.
25: **if** $f^\star(N) >$ best ▷ A better objective value found. **then**
26: best $= f^\star(N)$.
27: $N_{\text{optimal}} = N$.
28: **end if**
29: **end for**
30: Output - $d_1, \ldots, d_{N_{\text{optimal}}}$ from $x^\star(N_{\text{optimal}}), y^\star(N_{\text{optimal}})$ (2.32–2.36).
31: **end if**

Here, we have introduced the notation $z(N) = \min_{m\in\mathcal{M}} z_m(N)$. Now, a calculus-based argument identical to that in Lemma 2.6 from the single OAR case shows that $z_m(N)$ is an increasing sequence indexed by $N \ge 1$. Thus, $z(N)$ is also an increasing sequence. Moreover, an algebraic argument identical to that in

Lemma 2.5 shows that $D_m^\star(N) = Nd_m^\star(N)$ is bounded above by C_m/s_m, for all $N \geq 1$. This, in turn, implies by an algebraic argument identical to that in Lemma 2.6 that $z_m(N)$ is bounded above, for all $N \geq 1$. Thus, $z(N)$ is bounded above as well. This implies that $z(N)$ is a convergent sequence and that $f^\star(N)$ eventually decreases as claimed, by an argument identical to Lemma 2.6. Further, a calculus-based argument identical to that in the proof of Theorem 2.4 shows that the function $f_m^\star(N) = z_m(N) - \tau(N)$ is concave over all real numbers $N > 0$. Utilizing this in (3.42) shows that $f^\star(N)$ is the minimum of concave functions $f_m^\star(N)$ over $m \in \mathcal{M}$. Consequently, $f^\star(N)$ is concave over all real numbers $N > 0$. This implies that once this function drops from a higher value $f^\star(N^\star - 1)$ to a lower value $f^\star(N^\star)$, it cannot increase again. In other words, if we did not have an upper bound of $N_{\max}$ on the number of sessions, $N^\star - 1$ would be optimal. The formula $N_{\text{optimal}} = \min\{N^\star - 1, N_{\max}\}$ clips this optimal at the upper bound of $N_{\max}$, in view of the concavity of $f^\star(N)$. The optimal dose per session is then given by $c^\star(N_{\text{optimal}})$ as in Claim 1 of Proposition 3.1.

3.7 Numerical Experiments

Table 3.1 displays results obtained after applying Algorithm 1 to a problem with two OAR. All parameters for the tumor and the first OAR were identical to Table 2.1. Parameters for the second OAR were $T_2 = 35$ and $\delta_2 = 1.4286$ Gy per session. These are typical of brain stem in head-and-neck cancer. $N_{\max}$ was set at 100. The parameters in this example satisfied the sufficient condition for optimality of equal-dosage solutions. The table therefore reports the optimal number of sessions N_{optimal} and the dose $c^\star(N_{\text{optimal}})$ in Gy per session.

As in Table 2.1 for the single OAR case, N_{optimal} is increasing in T_{double} here as well. Table 3.1 sheds light on the complex trade-offs that arise between the tumor and the two OAR. As an illustrative scenario, consider the column of $T_{\text{double}} = 4$ days. The optimal number of sessions is 10 and the optimal dose per session is 3.38 Gy, when $s_1 = 1$ regardless of the values of s_2, ρ_2. Moreover, the optimal pair ($N_{\text{optimal}} = 10, c^\star(N_{\text{optimal}}) = 3.38$ Gy) is identical to that in Table 2.1, where the second OAR was not present. This suggests that, for this particular combination of parameter values, the BED constraint for the first OAR remains active with or without the presence of the second OAR. That is, the first OAR is the bottleneck. However, when $s_1 = 0.8$, the first OAR receives a smaller proportion of the dose administered to the tumor. Thus, the BED constraint on the first OAR becomes less

Table 3.1. ($N_{optimal}, c^{\star}(N_{optimal})$ *Gy) for different* $T_{double}, \rho_2, s_1, s_2$ *values when* ρ_1 *is fixed at* $1/5\ Gy^{-1}$.

		$\rho_1 = 1/5$ Gy^{-1}								
		$T_{\text{double}} = 2$ (days)			$T_{\text{double}} = 3$ (days)			$T_{\text{double}} = 4$ (days)		
		$s_1 = 1$	$s_1 = 0.8$	$s_1 = 0.6$	$s_1 = 1$	$s_1 = 0.8$	$s_1 = 0.6$	$s_1 = 1$	$s_1 = 0.8$	$s_1 = 0.6$
$\rho_2 = 1/5$ Gy^{-1}	$s_2 = 1$	(5, 5.43)	(5, 5.90)	(5, 5.90)	(7, 4.33)	(8, 4.31)	(8, 4.31)	(10, 3.38)	(11, 3.46)	(11, 3.46)
	$s_2 = 0.8$	(5, 5.43)	(4, 7.84)	(5, 7.37)	(7, 4.33)	(7, 5.41)	(8, 5.39)	(10, 3.38)	(9, 4.55)	(10, 4.62)
	$s_2 = 0.6$	(5, 5.43)	(4, 7.84)	(2, 16.09)	(7, 4.33)	(7, 5.41)	(4, 10.45)	(10, 3.38)	(9, 4.55)	(5, 9.04)
$\rho_2 = 1/4$ Gy^{-1}	$s_2 = 1$	(5, 5.43)	(7, 4.54)	(7, 4.54)	(7, 4.33)	(10, 3.58)	(10, 3.58)	(10, 3.38)	(14, 2.84)	(14, 2.84)
	$s_2 = 0.8$	(5, 5.43)	(4, 7.84)	(7, 5.68)	(7, 4.33)	(7, 5.41)	(11, 4.19)	(10, 3.38)	(9, 4.55)	(15, 3.38)
	$s_2 = 0.6$	(5, 5.43)	(4, 7.84)	(2, 16.09)	(7, 4.33)	(7, 5.41)	(4, 10.45)	(10, 3.38)	(9, 4.55)	(5, 9.04)
$\rho_2 = 1/3$ Gy^{-1}	$s_2 = 1$	(7, 4.32)	(8, 3.97)	(8, 3.97)	(8, 3.95)	(13, 2.89)	(13, 2.89)	(10, 3.38)	(18, 2.31)	(18, 2.31)
	$s_2 = 0.8$	(5, 5.43)	(7, 5.40)	(9, 4.60)	(7, 4.33)	(8, 4.94)	(15, 3.28)	(10, 3.38)	(9, 4.55)	(20, 2.69)
	$s_2 = 0.6$	(5, 5.43)	(4, 7.84)	(7, 7.20)	(7, 4.33)	(7, 5.41)	(7, 7.20)	(10, 3.38)	(9, 4.55)	(8, 6.58)

restrictive and the second OAR may therefore become the bottleneck. This indeed appears to be the case regardless of the value of ρ_2 when $s_2 = 1$ as reflected in the following observation. The optimal pairs ($N_{\text{optimal}} = 11, c^\star(N_{\text{optimal}}) = 3.46$ Gy), ($N_{\text{optimal}} = 14, c^\star(N_{\text{optimal}}) = 2.84$ Gy), and ($N_{\text{optimal}} = 18, c^\star(N_{\text{optimal}}) = 2.31$ Gy) are different from the optimal pair ($N_{\text{optimal}} = 9, c^\star(N_{\text{optimal}}) = 4.55$ Gy) that appears in Table 2.1 when the second OAR is absent. Interestingly, this effect disappears when s_2 reduces to 0.8 and 0.6, where the optimal pair reverts back to ($N_{\text{optimal}} = 9, c^\star(N_{\text{optimal}}) = 4.55$ Gy) – the same as when the second OAR is absent in Table 2.1. This is intuitive, because the second OAR receives a larger proportion of tumor dose (and hence is more likely to become the bottleneck) when $s_2 = 1$ as compared to when $s_2 = 0.8, 0.6$. A similar behavior is observed for $s_1 = 0.6$, but the effect there seems to be more conspicuous and more sensitive to the value of ρ_2. For instance, when $\rho_2 = 1/3$ Gy^{-1}, all three optimal pairs ($N_{\text{optimal}} = 18, c^\star(N_{\text{optimal}}) = 2.31$ Gy), ($N_{\text{optimal}} = 20, c^\star(N_{\text{optimal}}) = 2.69$ Gy), and ($N_{\text{optimal}} = 8, c^\star(N_{\text{optimal}}) = 6.58$ Gy) are different from the corresponding optimal pair ($N_{\text{optimal}} = 5, c^\star(N_{\text{optimal}}) = 9.04$ Gy) in Table 2.1 when the second OAR is absent. Similar patterns are also present, albeit at varying degrees of sensitivity, in the columns of $T_{\text{double}} = 2, 3$ days.

One limitation of formulations (P) and (Q) from the previous chapter and this chapter is that the values of the dose-response parameters in the LQ framework are difficult to estimate. The next two chapters propose two different approaches to tackle this challenge.

Bibliographic Notes

The material in this chapter is largely based on the doctoral dissertation [101] and associated papers [102, 103]. The algebraic proof in Section 3.1 seems to be new. Example 3.3 is from [103]. Table 3.1 is only the tip of the iceberg of the intricate trade-offs that arise in problem (Q). More extensive numerical results are reported in [103]. A brief overview of the results in this chapter and Algorithm 1 were included, without complete proofs, in a tutorial [61]. All results in this chapter can be extended to a model wherein tumor proliferation begins after an initial lag of T_{lag} days as mentioned in the Bibliographic Notes for Chapter 2. See [103] for detailed proofs that incorporate this proliferation lag. The equivalence between problem (Q(N)) and the two-variable problem (2VAR(N)) as in Lemma 3.2 was noted in [102]. Similarly for the equivalence between (2VAR(N)) and (2VARLP(N)) as in Lemma 3.4.

Exercises

Exercise 3.6 ⋆Attempt to construct an instance of problem (Q(N)) wherein an equal-dosage solution is optimal, but $\frac{\alpha_0}{\beta_0} < \max\limits_{m \in \mathcal{M}} \frac{1}{s_m} \frac{\alpha_m}{\beta_m}$. Similarly, attempt to construct an instance of problem (Q(N)) wherein a single-dosage solution is optimal, but $\frac{\alpha_0}{\beta_0} > \min\limits_{m \in \mathcal{M}} \frac{1}{s_m} \frac{\alpha_m}{\beta_m}$.

Exercise 3.7 Derive both optimal solutions reported in Example 3.3, using the methodology described in this chapter.

Exercise 3.8 ⋆Either prove or disprove by counterexample that the optimal objective values $f^\star(N)$ of problems (2VARLP(N)) are quasiconcave in N.

Exercise 3.9 Consider the dose-response parameters for the tumor and the two OAR given in Example 3.3. Compute the optimal number of treatment sessions and the corresponding optimal dose in each session, when $T_{\text{double}} \in \{2, 3, 4\}$ days.

4

Robust Fractionation

Since the values of dose-response parameters in the LQ framework are difficult to estimate, a dosing plan that is deemed to be optimal based on point estimates of these values may in fact be infeasible to the OAR constraints. In this chapter, we will study a robust optimization approach to address this concern.

4.1 Problem Formulation

While the treatment planner does not know the values of the OAR parameters $\rho_m = \beta_m/\alpha_m$, it assumes that these values belong to the intervals $[\rho_m^{\min}, \rho_m^{\max}]$, for $m \in \mathcal{M}$. Here, $\rho_m^{\min} > 0$ is the left end-point of this interval and $\rho_m^{\max} \geq \rho_m^{\min}$ is the right end-point. This is called an interval model of uncertainty or an interval model of ambiguity. The treatment planner is interested in dosing plans that remain feasible for all parameter values from these intervals. Among all such dosing plans, the treatment planner thus searches for one with the largest biological effect on the tumor. This worst-case approach is captured in the formulation

$$(\mathrm{R})\ g^{\star} = \max_{\vec{d},\, N}\ \alpha_0 \sum_{t=1}^{N} d_t + \beta_0 \sum_{t=1}^{N} (d_t)^2 - \tau(N) \tag{4.1}$$

$$s_m \sum_{t=1}^{N} d_t + s_m^2 \rho_m \sum_{t=1}^{N} (d_t)^2 \leq T_m \delta_m (1 + \rho_m \delta_m),$$

$$\forall \rho_m \in [\rho_m^{\min}, \rho_m^{\max}],\ m \in \mathcal{M} \tag{4.2}$$

$$\vec{d} \geq 0 \tag{4.3}$$

$$1 \leq N \leq N_{\max},\ \text{integer.} \tag{4.4}$$

This is called a robust counterpart of formulation (Q) from Chapter 3. We will not explicitly incorporate similar interval uncertainty in tumor parameters α_0, β_0 in the objective function. Since the objective function is increasing in these parameters, it would be optimal in the robust counterpart to simply fix α_0 and β_0 at their smallest values. Similarly, for the parameter T_{double}. The case of OAR parameters ρ_m, however, is complicated because those appear on both sides of (4.2). We remark that (4.2) in fact includes an uncountably infinite group of constraints. In this chapter, we will describe how these challenges can be resolved via a tailored solution.

As we did in Chapters 2 and 3, we will first focus on this problem with a fixed number of sessions $N \geq 1$. That is, we now consider

$$(\text{R}(N)) \; g^{\star}(N) = \max_{\vec{d}} \; \alpha_0 \sum_{t=1}^{N} d_t + \beta_0 \sum_{t=1}^{N} (d_t)^2 - \tau(N) \tag{4.5}$$

$$s_m \sum_{t=1}^{N} d_t + s_m^2 \rho_m \sum_{t=1}^{N} (d_t)^2 \leq T_m \delta_m (1 + \rho_m \delta_m),$$

$$\forall \rho_m \in [\rho_m^{\min}, \rho_m^{\max}], \; m \in \mathcal{M} \tag{4.6}$$

$$\vec{d} \geq 0. \tag{4.7}$$

We collect both ρ_m terms on the left-hand side and rewrite constraint (4.6) for OAR m as

$$s_m \sum_{t=1}^{N} d_t + \rho_m \left(s_m^2 \sum_{t=1}^{N} (d_t)^2 - T_m \delta_m^2 \right) \leq T_m \delta_m, \; \forall \rho_m \in [\rho_m^{\min}, \rho_m^{\max}]. \tag{4.8}$$

Dosing plans wherein $\sum_{t=1}^{N} (d_t)^2 > \frac{T_m \delta_m^2}{s_m^2}$ are feasible to (4.8) if, and only if,

$$s_m \sum_{t=1}^{N} d_t + \rho_m^{\max} \left(s_m^2 \sum_{t=1}^{N} (d_t)^2 - T_m \delta_m^2 \right) \leq T_m \delta_m. \tag{4.9}$$

To see why, notice that constraint (4.9) ensures that constraint (4.8) is satisfied for the largest value of the left-hand side over all values of the unknown parameter ρ_m from the interval $[\rho_m^{\min}, \rho_m^{\max}]$. Note, as an aside, that this largest value was obtained by using the right end-point $\rho_m^{\max}$ of the interval as the value of the unknown parameter ρ_m. This implies that constraint (4.8) is satisfied. Conversely, if constraint (4.8) is satisfied, then constraint (4.9) must be satisfied. Similarly, dosing plans wherein $\sum_{t=1}^{N} (d_t)^2 < \frac{T_m \delta_m^2}{s_m^2}$ are feasible to (4.8) if, and only if,

$$s_m \sum_{t=1}^{N} d_t + \rho_m^{\min} \left(s_m^2 \sum_{t=1}^{N} (d_t)^2 - T_m \delta_m^2 \right) \leq T_m \delta_m. \tag{4.10}$$

For dosing plans where $\sum_{t=1}^{N}(d_t)^2 = \frac{T_m \delta_m^2}{s_m^2}$, (4.8) reduces to $s_m \sum_{t=1}^{N} d_t \leq T_m \delta_m$. We employ these observations to solve (R(N)) by instead solving a set of $M+1$ subproblems as explained next.

4.2 Solution via a Finite Set of Subproblems

Define the notation $\mu_k = \frac{T_k \delta_k^2}{s_k^2}$, for $k = 1{:}\ M$. Suppose, without loss of generality, that the OAR are ordered such that $\mu_1 \leq \mu_2 \leq \cdots \leq \mu_M$. Then, in the kth subproblem, we restrict $\sum_{t=1}^{N}(d_t)^2$ as $\mu_{k-1} \leq \sum_{t=1}^{N}(d_t)^2 \leq \mu_k$. We do this for $k = 1, 2, \ldots, M+1$, with the convention that $\mu_0 = 0$ and $\mu_{M+1} = \infty$. Thus, in the kth subproblem, constraint (4.6) is replaced with constraint (4.9) for OAR $m < k$, whereas it is replaced with (4.10) for OAR $m \geq k$. Thus, the kth subproblem is formulated as

$$(\mathrm{R}(N,k))\ g^\star(N,k) = \max_{\vec{d}}\ \alpha_0 \sum_{t=1}^{N} d_t + \beta_0 \sum_{t=1}^{N} (d_t)^2 - \tau(N)$$

$$s_m \sum_{t=1}^{N} d_t + \rho_m^{\max} s_m^2 \sum_{t=1}^{N} (d_t)^2 \leq T_m \delta_m (1 + \rho_m^{\max} \delta_m),\ m = 1{:}\ k-1$$

$$s_m \sum_{t=1}^{N} d_t + \rho_m^{\min} s_m^2 \sum_{t=1}^{N} (d_t)^2 \leq T_m \delta_m (1 + \rho_m^{\min} \delta_m),\ m = k{:}\ M$$

$$\mu_{k-1} \leq \sum_{t=1}^{N} (d_t)^2 \leq \mu_k \tag{4.11}$$

$$\vec{d} \geq 0.$$

Similar to (2VAR(N)) from Lemma 3.2, the above subproblem can be reformulated as the two-variable subproblem

$$(\mathrm{2VARROB}(N,k))\ g^\star(N,k) = \max_{x,y}\ \alpha_0 x + \beta_0 y - \tau(N),$$

$$s_m x + \rho_m^{\max} s_m^2 y \leq T_m \delta_m (1 + \rho_m^{\max} \delta_m),\ m = 1{:}\ k-1$$

$$s_m x + \rho_m^{\min} s_m^2 y \leq T_m \delta_m (1 + \rho_m^{\min} \delta_m),\ m = k{:}\ M$$

$$\mu_{k-1} \leq y \leq \mu_k \tag{4.12}$$

$$\sqrt{y} \leq x \leq \sqrt{Ny}$$

$$x \geq 0, y \geq 0.$$

This problem has the same form as (2VAR(N)) except for constraints (4.12). These complicating bound constraints can be tackled by considering four mutually exclusive and exhaustive cases as described next.

Let γ_k° denote the best single-dosage solution in (R(N,k)) without the bound constraints (4.11), calculated similar to (3.11). Similarly, let $c_k^\star(N)$ denote the best equal dose per session in (R(N,k)) without the bound constraints (4.11), calculated similar to (3.9). The four cases are rooted in distinct possibilities for the relative values of $\mu_k, \mu_{k-1}, (\gamma_k^\circ)^2$, and $N(c_k^\star(N))^2$, as shown in Figure 4.1. The feasible region of (2VARROB(N,k)) without the bounding constraints (4.12) has the same structure as that in (2VAR(N)) (recall Figure 3.1). This is shown as a gray horn-shaped area. The bounding constraints (4.12) are depicted as thick black horizontal lines at μ_k (solid) and μ_{k-1} (dash-dotted). Point A corresponds to the best single-dosage solution, where $y = (\gamma_k^\circ)^2$ is shown as a thin dashed horizontal line. Point B corresponds

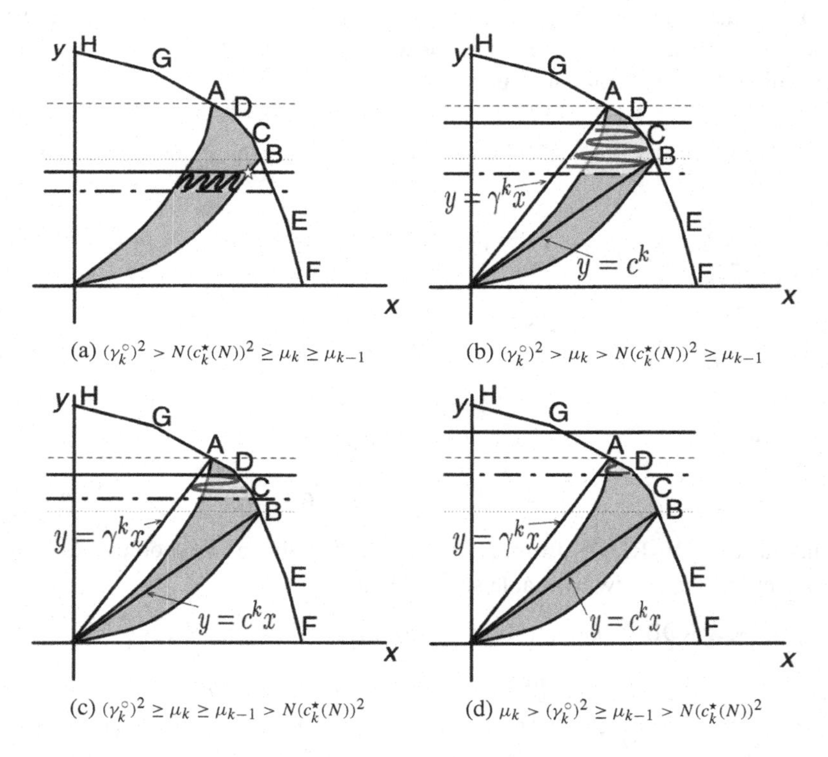

(a) $(\gamma_k^\circ)^2 > N(c_k^\star(N))^2 \geq \mu_k \geq \mu_{k-1}$

(b) $(\gamma_k^\circ)^2 > \mu_k > N(c_k^\star(N))^2 \geq \mu_{k-1}$

(c) $(\gamma_k^\circ)^2 \geq \mu_k \geq \mu_{k-1} > N(c_k^\star(N))^2$

(d) $\mu_k > (\gamma_k^\circ)^2 \geq \mu_{k-1} > N(c_k^\star(N))^2$

Figure 4.1 Four subcases that facilitate exact solution of (2VARROB(N,k)). In each case, the feasible region of the problem that is actually solved is hatched with a squiggly pattern. Reused with permission from [61, figure 9].

to the best equal-dosage solution, where $y = N(c_k^\star(N))^2$ is shown as a thin dotted horizontal line.

In the first case, an optimal solution occurs at the intersection of $x = \sqrt{N\mu_k}$ and $y = \mu_k$. This is shown in Figure 4.1(a) as a star. In the other three cases, the nonconvex feasible region can be expanded to a polygon without altering optimal solutions (see Figures 4.1(b)–(d)). This follows by a reasoning similar to problem (2VARLP(N)) as was explained in Figure 3.2. In each of these three cases, an optimal solution can thus be found by solving the two-variable LP

$$
\begin{aligned}
(\text{2VARROBLP}(N,k))\ \max_{x,y}\ & \alpha_0 x + \beta_0 y - \tau(N),\\
s_m x + \rho_m^{\max} s_m^2 y &\le T_m \delta_m (1 + \rho_m^{\max} \delta_m),\ m = 1:k-1\\
s_m x + \rho_m^{\min} s_m^2 y &\le T_m \delta_m (1 + \rho_m^{\min} \delta_m),\ m = k:M\\
& \mu_{k-1} \le y \le \mu_k\\
& c_k^\star(N) x \le y\\
& y \le \gamma_k^\circ x\\
& x \ge 0, y \ge 0.
\end{aligned}
$$

A closer look at Figures 4.1(b)–(d) provides further insight into the structure of the feasible region of this LP. In the first case (Figure 4.1(b)), none of the constraints are redundant; in the second case (Figure 4.1(c)), the constraint $c_k^\star(N)x \le y$ is redundant, whereas in the third case (Figure 4.1(d)), constraints $c_k^\star(N)x \le y$ and $y \le \mu_k$ are redundant.

In summary, subproblem (2VARROB(N,k)) calls for the solution of at most one two-variable LP. This yields the largest objective value $g^\star(N,k)$ attained at the pair $(x^\star(N,k), y^\star(N,k))$. Then, comparing these $g^\star(N,k)$ values across subproblems $k = 1: M+1$ yields the largest objective value $g^\star(N)$ of (R(N)). Suppose this is attained in the subproblem indexed $k^\star(N)$. Then, comparing the values of $g^\star(N)$ across $N = 1: N_{\max}$ yields the largest objective value $g^\star$ in problem (R). Suppose this is attained when the number of sessions is N_{optimal}. Consider the pair $(x^\star(N_{\text{optimal}}, k^\star(N_{\text{optimal}})), y^\star(N_{\text{optimal}}, k^\star(N_{\text{optimal}})))$. Figures 4.1(a)–(d) collectively show that no matter which case this optimal pair came from, it will satisfy the relationship $\sqrt{y} \le x \le \sqrt{Ny}$. Thus, a dosing plan that is optimal to (R) can be recovered from

$$(x^\star(N_{\text{optimal}}, k^\star(N_{\text{optimal}})), y^\star(N_{\text{optimal}}, k^\star(N_{\text{optimal}})))$$

as explained in Lemma 2.2 from Chapter 2. The overall procedure is outlined in Algorithm 2.

Algorithm 2 Exact solution of (R)

1: Input - $\alpha_0, \beta_0, T_{\text{double}}$; set $\mathcal{M}$ of OAR; $T_m, \delta_m, \rho_m^{\min}, \rho_m^{\max}, s_m$, for all $m \in \mathcal{M}$; $N_{\max}$.
2: Calculate $\mu_m = \frac{T_m \delta_m^2}{s_m^2}$, for all $m \in \mathcal{M}$.
3: Sort the OAR indices $\{1,2,\ldots,m\}$ such that $\mu_1 \leq \mu_2 \leq \ldots \leq \mu_M$.
4: Let $\mu_0 = 0$ and $\mu_{M+1} = \infty$.
5: Reorder $T_m, \delta_m, \rho_m^{\min}, \rho_m^{\max}, s_m$ according to the sorted OAR indices.
6: **for** $N = 1 : N_{\max}$ **do**
7: **for** $k = 1 : M+1$ **do**
8: Let $\rho_m = \rho_m^{\max}$, for $m = 1 : k-1$; and $\rho_m = \rho_m^{\min}$, for $m = k : M$.
9: Let $C_m = T_m \delta_m (1 + \rho_m \delta_m)$, for $m = 1 : M$.
10: Calculate $c_k^\star(N)$ from (3.9).
11: Calculate γ_k° from (3.11).
12: **if** $(\gamma_k^\circ)^2 > N(c_k^\star(N))^2 \geq \mu_k$ ▷ Figure 4.1(a). **then**
13: Let $x^\star(N,k) = \sqrt{N\mu_k}$ and $y^\star(N,k) = \mu_k$.
14: **else** ▷ Figure 4.1(b)–(d).
15: Find $(x^\star(N,k), y^\star(N,k))$ optimal to (2VARROBLP(N,k)).
16: **end if**
17: Calculate $g^\star(N,k) = \alpha_0 x^\star(N,k) + \beta_0 y^\star(N,k) - \tau(N)$.
18: **end for**
19: Calculate $g^\star(N) = \max_{k=1:M+1} g^\star(N,k)$ and $k^\star(N) \in \operatorname{argmax}_{k=1:M+1} g^\star(N,k)$.
20: **end for**
21: Calculate $g^\star = \max_{N=1:N_{\max}} g^\star(N)$ and $N_{\text{optimal}} \in \operatorname{argmax}_{N=1:N_{\max}} g^\star(N)$.
22: Output - $d_1, \ldots, d_{N_{\text{optimal}}}$ using (2.32–2.36) from $x^\star(N_{\text{optimal}}, k^\star(N_{\text{optimal}})), y^\star(N_{\text{optimal}}, k^\star(N_{\text{optimal}}))$.

4.3 Numerical Experiments

We consider an example with two OAR and data as in Section 3.7. Parameters for the first OAR were fixed at $T_1 = 35$, $s_1 = 1$, and $\delta_1 = 1.2857$ Gy per session. Parameters for the second OAR were fixed at $T_2 = 35$, $s_2 = 1$, and $\delta_2 = 1.4286$ Gy per session. Interval uncertainty sets were constructed as follows. We fixed nominal values of OAR dose-response parameters at $\rho_1 = 1/5 = 0.2$ Gy^{-1} and $\rho_2 = 1/4 = 0.25$ Gy^{-1}. Note that these values are unknown to the treatment planner and are not utilized anywhere by our solution algorithm. They were employed in building uncertainty intervals with $\rho_1^{\min} = (1-\Delta)\rho_1$, $\rho_1^{\max} = (1+\Delta)\rho_1$ for the first OAR and $\rho_2^{\min} = (1-\Delta)\rho_2$, $\rho_2^{\max} = (1+\Delta)\rho_2$ for the second OAR. We used $\Delta \in \{0, 0.1, 0.2, \ldots, 0.9\}$ in our *parameters, for $\alpha_0/\beta_0 = 10$ Gy (top three panels) and $\alpha_0/\beta_0 = 20$ Gy (bottom three panels) across distinct T_{double} values. The optimal number of sessions (columns labeled $N_{optimal}$) and the optimal dose administered in each of these sessions (columns labeled dose) are also listed; an equal-dosage solution was optimal in each case.*

	$\alpha_0 = 0.35$ Gy^{-1}, $\beta_0 = 0.035$ Gy^{-1}								
	$T_{double} = 2$ (days)			$T_{double} = 3$ (days)			$T_{double} = 4$ (days)		
Δ	$N_{optimal}$	dose (Gy)	%	$N_{optimal}$	dose (Gy)	%	$N_{optimal}$	dose (Gy)	%
0	5	5.42	0	7	4.33	0	10	3.38	0
0.1	5	5.32	2.91	8	3.89	2.26	11	3.11	1.81
0.2	6	4.64	5.43	9	3.54	4.18	12	2.89	3.33
0.3	6	4.57	7.56	10	3.25	5.85	13	2.71	4.62
0.4	6	4.51	9.48	10	3.22	7.28	14	2.55	5.74
0.5	7	4.03	11.17	11	2.99	8.56	14	2.53	6.72
0.6	7	3.99	12.66	11	2.96	9.68	15	2.39	7.57
0.7	7	3.94	14.02	12	2.77	10.71	16	2.28	8.33
0.8	7	3.91	15.28	12	2.75	11.62	16	2.26	9.01
0.9	8	3.56	16.40	12	2.74	12.46	17	2.16	9.61
	$\alpha_0 = 0.70$ Gy^{-1}, $\beta_0 = 0.035$ Gy^{-1}								
	$T_{double} = 2$ (days)			$T_{double} = 3$ (days)			$T_{double} = 4$ (days)		
Δ	$N_{optimal}$	dose (Gy)	%	$N_{optimal}$	dose (Gy)	%	$N_{optimal}$	dose (Gy)	%
0	14	2.64	0	20	2.01	0	26	1.64	0
0.1	15	2.49	1.25	22	1.86	0.72	27	1.58	0.38
0.2	16	2.35	2.32	23	1.79	1.30	29	1.49	0.66
0.3	17	2.23	3.27	24	1.73	1.79	30	1.45	0.85
0.4	17	2.22	4.1	25	1.67	2.20	32	1.38	0.98
0.5	18	2.12	4.83	26	1.61	2.54	33	1.35	1.06
0.6	19	2.03	5.49	27	1.57	2.81	35	1.28	1.09
0.7	19	2.02	6.07	28	1.52	3.04	35	1.28	1.09
0.8	20	1.94	6.60	28	1.52	3.04	35	1.28	1.09
0.9	20	1.93	7.07	30	1.44	3.38	35	1.28	1.09

experiments. Here, $\Delta = 0$ recovers the nominal problem with no uncertainty, and the larger the Δ, the larger the uncertainty interval. We implemented Algorithm 2 on this example. Tumor parameters were set at $\beta_0 = 0.035$ Gy^{-2}, $\alpha_0 \in \{0.35, 0.70\}$ Gy^{-1}, and $T_{\text{double}} \in \{2, 3, 4\}$ days similar to Section 2.7. The results are displayed in Table 4.1.

The table shows that when $\alpha_0 = 0.35$ Gy^{-1}, the optimal number of sessions and the dose per session in the row of $\Delta = 0$ match with the corresponding cells from Table 3.1. This is expected, because when $\Delta = 0$, the uncertainty interval is a singleton, and thus the robust problem reduces to the nominal problem as there is no uncertainty in the values of OAR dose parameters. In each panel of the table, the percentage price of robustness for each value of Δ was calculated as the percentage relative difference between the optimal objective values of the nominal problem (that is, when $\Delta = 0$) and the robust problem. That is, 100(optimal nominal objective-optimal robust objective)/(optimal nominal objective). Thus, this percentage price of robustness is 0 when $\Delta = 0$ throughout the table. The table shows that the percentage price of robustness increases as Δ increases. This is intuitive because the size of interval uncertainty sets increases with Δ, thus rendering constraint (4.2) more restrictive. That is, the optimal objective value of the robust problem decreases with increasing Δ. Other qualitative trends can also be observed in the table. For instance, the optimal number of sessions increases and the optimal dose per session decreases with increasing values of Δ. There does not appear to be a straightforward explanation for this, and in fact, the trend might not hold for larger values of T_{double}. The percentage price of robustness is smaller for $\alpha_0 = 0.7$ Gy^{-1} than for $\alpha_0 = 0.35$ Gy^{-1}, for each fixed value of $\Delta \neq 0$. Other qualitative trends noted in Sections 2.7 and 3.7 for the nominal problem are also present in the robust case here. For example, at each fixed value of Δ, the optimal number of sessions increases with increasing values of T_{double} because the tumor growth is slower.

In the next chapter, we will study an alternative approach to incorporate uncertainty in dose-response parameters.

Bibliographic Notes

The methodology proposed in this chapter is similar to the doctoral dissertation [4]. A similar analysis was also independently performed in [13]. A brief summary of the material here was included in a tutorial [61], without listing Algorithm 2 and the accompanying numerical results. Extensive numerical experiments and sensitivity analyses are available in the doctoral dissertation [4]. We refer readers to the textbook [14] for methodological details about robust optimization. Survey papers on robust optimization include [19] and [55].

Exercises

Exercise 4.1 *Either prove or disprove by counterexample that the optimal objective values $g^\star(N)$ of problems (R(N)) are quasiconcave in N.

Exercise 4.2 Observe in Table 4.1 that when $\alpha_0 = 0.7$ Gy^{-1} and $T_{\text{double}} = 4$ days, the optimal number of sessions remains constant at 35 and the percentage price of robustness also remains fixed at 1.09, even while Δ increases as $0.6, 0.7, 0.8, 0.9$. Attempt to explain this peculiar behavior.

Exercise 4.3 Explore whether (and in what way) the percentage price of robustness is sensitive to the nominal value of OAR dose-response parameter ρ_1. Fix all other parameters at values given in Section 4.3.

Exercise 4.4 Design a numerical experiment to explore the amount and frequency of constraint violations incurred by the robust and the nominal optimal solutions, when the realized value of dose-response parameter ρ_1 lies outside the assumed interval of uncertainty. Refer to section 3.3 from Ajdari and Ghate [6] for ideas.

Exercise 4.5 *Explore whether or not you can construct an example where an optimal solution to problem (2VARROB(N,k)) corresponds to an unequal-dosage plan.

Exercise 4.6 Suppose that the OAR sparing factors s_m, for $m \in \mathcal{M}$, are also unknown, in addition to the OAR dose-response parameters. Suppose the treatment planner assumes that they belong to intervals $[s_m^{\min}, s_m^{\max}]$, with $0 < s_m^{\min} \leq s_m^{\max}$. Discuss how you will extend the robust optimization approach proposed in the chapter to tackle this situation.

5
Inverse Fractionation

Inverse optimization involves inferring the "best" values of the parameters of an optimization model, using given values of decision variables. This is in contrast to the usual (forward) optimization, where the values of decision variables are determined using given parameter values.

Since the dose-response parameters in the LQ framework are unknown and difficult to estimate, we discuss an inverse optimization approach. In particular, we infer the dose-response parameters that would render a given dosing plan optimal. Mathematically, we recall problem (Q(N)) from Chapter 3 and suppose that a dosing plan $\vec{d}^\star = (d_1^\star, \ldots, d_N^\star)$ is available to the treatment planner. We then wish to find tumor LQ dose-response parameters $\alpha_0 > 0, \beta_0 > 0$, and OAR dose-response parameters $\rho_m > 0$, for all $m \in \mathcal{M}$, that would render $\vec{d}^\star$ optimal to (Q(N)). We will describe a procedure that provides closed-form formulas for all possible values of such parameters.

Since N is fixed, $\tau(N)$ is a constant and has no effect on our analysis; it is thus dropped from further discussion. Let $x^\star = \sum_{t=1}^N d_t^\star$ and $y^\star = \sum_{t=1}^N (d_t^\star)^2$. Recall from Lemma 3.2 in Chapter 3 that problems (Q(N)) and (2VAR(N)) are equivalent. That is, the dosing plan $\vec{d}^\star$ is optimal to (Q(N)) if, and only if, the pair $(x^\star, y^\star)$ is optimal to (2VAR(N)). Thus, we work with the pair $(x^\star, y^\star)$ and the problem (2VAR(N)). As such, we describe a procedure that provides closed-form formulas for all possible values of parameters $\alpha_0 > 0, \beta_0 > 0$, and ρ_m, for $m \in \mathcal{M}$, that render $(x^\star, y^\star)$ optimal to (2VAR(N)).

5.1 Necessary, Sufficient Conditions for Feasibility and Optimality

We begin with an important lemma that is easy to prove.

Lemma 5.1 *There exists a $\rho_m > 0$ that makes $(x^\star, y^\star)$ feasible to constraint* (3.30) *in (2VAR(N)) for OAR $m \in \mathcal{M}$ if, and only if, one of the following six* *mutually exclusive conditions holds:*

$$s_m x^\star = T_m \delta_m \text{ and } s_m^2 y^\star = T_m \delta_m^2 \tag{5.1}$$

$$s_m x^\star < T_m \delta_m \text{ and } s_m^2 y^\star = T_m \delta_m^2 \tag{5.2}$$

$$s_m x^\star = T_m \delta_m \text{ and } s_m^2 y^\star < T_m \delta_m^2 \tag{5.3}$$

$$s_m x^\star < T_m \delta_m \text{ and } s_m^2 y^\star < T_m \delta_m^2 \tag{5.4}$$

$$s_m x^\star < T_m \delta_m \text{ and } s_m^2 y^\star > T_m \delta_m^2 \tag{5.5}$$

$$s_m x^\star > T_m \delta_m \text{ and } s_m^2 y^\star < T_m \delta_m^2. \tag{5.6}$$

Proof We first recall that the constant C_m that appears on the right-hand side of constraint (3.30) is our shorthand for $T_m \delta_m (1 + \rho_m \delta_m)$.
To establish the "if" part, we consider the six conditions in two groups.
Suppose any one of (5.1)–(5.4) holds. Then, for any $\rho_m > 0$,

$$(s_m x^\star - T\delta_m) + \rho_m (s_m^2 y^\star - T\delta_m^2) \leq 0.$$

That is, constraint (3.30) is satisfied.
Suppose condition (5.5) or condition (5.6) holds. Let $\rho_m = \frac{T\delta_m - s_m x^\star}{s_m^2 y^\star - T\delta_m^2} > 0$. The strict positivity here holds because the numerator and the denominator are not zero but have the same sign. Then,

$$\begin{aligned}
&(s_m x^\star - T\delta_m) + \rho_m (s_m^2 y^\star - T\delta_m^2) \\
&= (s_m x^\star - T\delta_m) + \frac{T\delta_m - s_m x^\star}{s_m^2 y^\star - T\delta_m^2}(s_m^2 y^\star - T\delta_m^2) = 0.
\end{aligned}$$

That is, constraint (3.30) is satisfied.
This shows constructively that if any one of (5.1)–(5.6) holds, then there exists a $\rho_m > 0$ such that constraint (3.30) is satisfied, for OAR $m \in \mathcal{M}$. For the "only if" part, suppose none of the six conditions holds for OAR m. Then, the only remaining possibilities are that either $s_m x^\star = T\delta_m$ and $s_m^2 y^\star > T\delta_m^2$ or $s_m x^\star > T\delta_m$ and $s_m^2 y^\star \geq T\delta_m^2$. Then, $(s_m x^\star - T\delta_m) + \rho_m (s_m^2 y^\star - T\delta_m^2) > 0$, for every $\rho_m > 0$. That is, there is no $\rho_m > 0$ that renders $x^\star, y^\star$ feasible to constraint (3.30). This completes the proof. □

Informally, this lemma implies, in other words, that if there were an OAR $m \in \mathcal{M}$ that does not satisfy any of the conditions (5.1)–(5.6), then there is no $\rho_m > 0$ that would render $(x^\star, y^\star)$ even feasible to problem (2VAR(N)). Thus, the inverse optimization problem cannot be solved, that is, it is ill-posed, unless the given dosing plan satisfies the following assumption.

Assumption 5.2 (Feasibility) *Exactly one of the six conditions* (5.1)–(5.6) *holds for each OAR* $m \in \mathcal{M}$.

We denote the set of OAR that satisfy condition (5.1) by $\mathcal{M}_1$; the set of OAR that satisfy condition (5.2) by $\mathcal{M}_2$; and so on, up to $\mathcal{M}_6$ for condition (5.6). We know that $\mathcal{M}_i \bigcap \mathcal{M}_j$, for any pair of OAR $i \neq j$, by definition. Assumption 5.2 requires that these six sets be exhaustive. That is, we must have that $\mathcal{M} = \bigcup_{m=1}^{6} \mathcal{M}_m$.

Condition (5.1) implies that constraint (3.30) is active at $x^\star, y^\star$, regardless of the value of $\rho_m > 0$, for each OAR $m \in \mathcal{M}_1$. Condition (5.2) implies that constraint (3.30) is inactive at $x^\star, y^\star$, regardless of the value of $\rho_m > 0$, for each OAR $m \in \mathcal{M}_2$. Similarly, condition (5.3) shows that constraint (3.30) is inactive at $x^\star, y^\star$, regardless of the value of $\rho_m > 0$, for each OAR $m \in \mathcal{M}_3$. Finally, condition (5.4) shows that constraint (3.30) is inactive at $x^\star, y^\star$, regardless of the value of $\rho_m > 0$, for each OAR $m \in \mathcal{M}_4$. Thus, values of $\rho_m > 0$, for $m \in \mathcal{M}_1 \cup \mathcal{M}_2 \cup \mathcal{M}_3 \cup \mathcal{M}_4$ can be fixed arbitrarily. We will later see, however, that the value of ρ_m, for $m \in \mathcal{M}_1$, may affect the tumor parameters $\alpha_0 > 0$ and $\beta_0 > 0$ that render the given dosing plan optimal. This is not the case for $\mathcal{M}_2, \mathcal{M}_3, \mathcal{M}_4$. It is for this reason that we will simply say that $\rho_m \in (0, \infty)$, for $m \in \mathcal{M}_2 \cup \mathcal{M}_3 \cup \mathcal{M}_4$.

Lemma 5.3 *Suppose $\mathcal{M}_1 \cup \mathcal{M}_5 \cup \mathcal{M}_6 = \emptyset$. Then, there are no $\alpha_0 > 0$ and $\beta_0 > 0$ for which $x^\star, y^\star$ can be optimal to (2VAR(N)).*

Proof Recall that there is no $\rho_m > 0$ for which constraint (3.30) can be active at $x^\star, y^\star$, for any OAR $m \in \mathcal{M}_2 \cup \mathcal{M}_3 \cup \mathcal{M}_4$. Thus, $x^\star, y^\star$ cannot lie on a "frontier" such as ADCB as schematically depicted in Figure 3.1 from Chapter 3. That is, there are no $\alpha_0 > 0$ and $\beta_0 > 0$ for which $x^\star, y^\star$ can be optimal to (2VAR(N)). □

Thus, it is essential to make the following assumption; without this assumption, the inverse problem does not have a solution.

Assumption 5.4 (Optimality) $\mathcal{M}_1 \cup \mathcal{M}_5 \cup \mathcal{M}_6 \neq \emptyset$.

We constructively demonstrate next that this assumption is also sufficient for rendering $(x^\star, y^\star)$ optimal to (2VAR(N)). The construction is based on the observation (recall Figure 3.1) that at least one among constraints (3.30) must be active at an optimal solution to (2VAR(N)).

5.2 Closed-Form Formulas for Imputed Parameter Values

The idea is to identify all possible combinations of OAR for which constraint (3.30) could be active at $(x^\star, y^\star)$. Candidate OAR for this are from the set

$\mathcal{M}_1 \cup \mathcal{M}_5 \cup \mathcal{M}_6$ because, as described previously, constraint (3.30) can never be active for other OAR. OAR parameter values that would render each particular combination of constraints active at $(x^\star, y^\star)$ are then determined. Tumor parameter values that would induce an optimal solution to occur at that particular combination of active constraints at $(x^\star, y^\star)$ are then identified. This recovers all possible combinations of OAR and tumor parameter values that make $(x^\star, y^\star)$ optimal. As we show next, the analysis is based on calculations and comparisons of slopes of various linear constraints and objective function contours. The analysis becomes easier to describe by first writing the linear constraint (3.30) in its expanded form

$$s_m x + s_m^2 \rho_m y \leq T\delta_m(1 + \rho_m \delta_m),\ m \in \mathcal{M}, \tag{5.7}$$

and observing that the slope of its boundary is $-1/(\rho_m s_m) < 0$.

5.2.1 Slopes of Constraint Boundaries for OAR in $\mathcal{M}_1$

Recall that constraint (5.7) is active at $(x^\star, y^\star)$, for each OAR in $m \in \mathcal{M}_1$, regardless of the (now fixed) value of $\rho_m > 0$. We define

$$\mu_m = \frac{-1}{\rho_m s_m} < 0,\ m \in \mathcal{M}_1, \tag{5.8}$$

as the slopes of the corresponding constraint boundaries. Also let

$$\underline{\mu} = \begin{cases} \min\limits_{m \in \mathcal{M}_1} -\mu_m, & \text{if } \mathcal{M}_1 \neq \emptyset \\ \infty \end{cases} \qquad \bar{\mu} = \begin{cases} \max\limits_{m \in \mathcal{M}_1} -\mu_m, & \text{if } \mathcal{M}_1 \neq \emptyset \\ -\infty \end{cases} \tag{5.9}$$

be the flattest and steepest slope magnitudes of constraint boundaries (5.7), over all $m \in \mathcal{M}_1$. Observe that if $\mathcal{M}_1$ is a singleton, then $\underline{\mu} = \bar{\mu}$.

5.2.2 Slopes of Constraint Boundaries for OAR in $\mathcal{M}_5 \cup \mathcal{M}_6$

We consider each subset of OAR in $\mathcal{M}_5$ and $\mathcal{M}_6$ and find $\rho_m > 0$ values that would make constraint (5.7) active at $(x^\star, y^\star)$ for those OAR. For all other OAR in $\mathcal{M}_5$ and $\mathcal{M}_6$, we identify $\rho_m > 0$ values that would make constraint (5.7) inactive at $(x^\star, y^\star)$. To facilitate discussion, let $\mathcal{M}_i^a$, for $i = 5, 6$, denote the (possibly empty) sets of OAR in $\mathcal{M}_i$ for which constraint (5.7) will be made active at $(x^\star, y^\star)$. If $\mathcal{M}_1$ is empty, then at least one of $\mathcal{M}_5^a$ and $\mathcal{M}_6^a$ must be nonempty. That is, $\mathcal{M}_5^a$ and $\mathcal{M}_6^a$ must be chosen such that

$$\mathcal{M}_1 \cup \mathcal{M}_5^a \cup \mathcal{M}_6^a \neq \emptyset. \tag{5.10}$$

Assumption 5.4 guarantees that this can be done.

We must have that $s_k x^\star + s_k^2 \rho_k y^\star < T\delta_k(1 + \rho_k \delta_k)$, for all OAR $k \in \{\mathcal{M}_5 \setminus \mathcal{M}_5^a\} \cup \{\mathcal{M}_6 \setminus \mathcal{M}_6^a\}$, because we want to render constraint (5.7) inactive at $(x^\star, y^\star)$ for these OAR. That is, $\rho_k(s_k^2 y^\star - T\delta_k^2) < T\delta_k - s_k x^\star$ for all such k. Then, suitable values of ρ_k are

$$0 < \rho_k < \frac{(T\delta_k - s_k x^\star)}{(s_k^2 y^\star - T\delta_k^2)}, \text{ if } k \in \mathcal{M}_5 \setminus \mathcal{M}_5^a \tag{5.11}$$

$$\rho_k > \frac{(T\delta_k - s_k x^\star)}{(s_k^2 y^\star - T\delta_k^2)}, \text{ if } k \in \mathcal{M}_6 \setminus \mathcal{M}_6^a. \tag{5.12}$$

Conditions (5.5)–(5.6) guarantee that these upper and lower bounds on ρ_k are both positive, for all $k \in \mathcal{M}_5 \setminus \mathcal{M}_5^a$ and $k \in \mathcal{M}_6 \setminus \mathcal{M}_6^a$, respectively. Now let $\mathcal{M}^a = \mathcal{M}_5^a \cup \mathcal{M}_6^a$ denote the subset of OAR from $\mathcal{M}_5 \cup \mathcal{M}_6$ for which we want to make constraint (5.7) active at $(x^\star, y^\star)$. Thus, $s_m x^\star + s_m^2 \rho_m y^\star = T\delta_m(1 + \rho_m \delta_m)$ must hold for all $m \in \mathcal{M}^a$. The resulting value of ρ_m is given by

$$\rho_m = \frac{T\delta_m - s_m x^\star}{s_m^2 y^\star - T\delta_m^2}, \ \forall m \in \mathcal{M}^a. \tag{5.13}$$

Conditions (5.5)–(5.6) guarantee that this $\rho_m > 0$, for all $m \in \mathcal{M}^a$. The equation of the boundary of any active constraint in $\mathcal{M}^a$ is

$$y = -\frac{1}{s_m}\left(\frac{s_m^2 y^\star - T\delta_m^2}{T\delta_m - s_m x^\star}\right) x + \frac{T\delta_m}{s_m}\left(\frac{s_m y^\star - \delta_m x^\star}{T\delta_m - s_m x^\star}\right).$$

This equation was obtained by rearranging terms from (5.7). The slope of this linear constraint boundary is

$$\sigma_m = -\frac{1}{s_m}\left(\frac{s_m^2 y^\star - T\delta_m^2}{T\delta_m - s_m x^\star}\right) < 0. \tag{5.14}$$

Let

$$\underline{\sigma} = \begin{cases} \min\limits_{m \in \mathcal{M}^a} -\sigma_m, & \text{if } \mathcal{M}^a \neq \emptyset \\ \infty \end{cases} \qquad \bar{\sigma} = \begin{cases} \max\limits_{m \in \mathcal{M}^a} -\sigma_m, & \text{if } \mathcal{M}^a \neq \emptyset \\ -\infty \end{cases} \tag{5.15}$$

be the flattest and steepest slope magnitudes of constraint boundaries (5.7) over all $\mathcal{M}^a$. Observe that if $\mathcal{M}^a$ is a singleton, then $\underline{\sigma} = \bar{\sigma}$.

Now let

$$L = \min\{\underline{\sigma}, \underline{\tau}\}, \text{ and } U = \max\{\bar{\sigma}, \bar{\tau}\} \tag{5.16}$$

be the flattest and steepest slope magnitudes of constraint boundaries (5.7) over all active constraints. Equation (5.10) guarantees that both L and U are finite. This helps determine suitable values of $\alpha_0 > 0$ and $\beta_0 > 0$ via three separate cases.

5.2.3 Three Cases to Impute Tumor Parameters

Case 1 (Figure 5.1) Constraint (3.31) is active at optimal solution $(x^\star, y^\star)$; that is, $x^\star = \sqrt{y^\star}$.

The figure shows that the magnitude α_0/β_0 of the optimal objective contour slope must be no larger than the magnitude U of the slope of the steepest active constraint. Thus, any $\alpha_0 > 0$ and $\beta_0 > 0$ values with

$$\frac{\alpha_0}{\beta_0} \leq U \tag{5.17}$$

are suitable.

Case 2 (Figure 5.2) Constraint (3.32) is active at optimal solution $(x^\star, y^\star)$; that is, $x^\star = \sqrt{Ny^\star}$.

The figure shows that the magnitude α_0/β_0 of the optimal objective contour slope must be no smaller than the magnitude of the slope of the flattest active constraint. Thus, any $\alpha_0 > 0$ and $\beta_0 > 0$ values with

$$\frac{\alpha_0}{\beta_0} \geq L \tag{5.18}$$

are suitable.

Case 3 (Figure 5.3) Neither constraint (3.31) nor constraint (3.32) is active at optimal solution $(x^\star, y^\star)$; that is, $\sqrt{y^\star} < x^\star < \sqrt{Ny^\star}$.

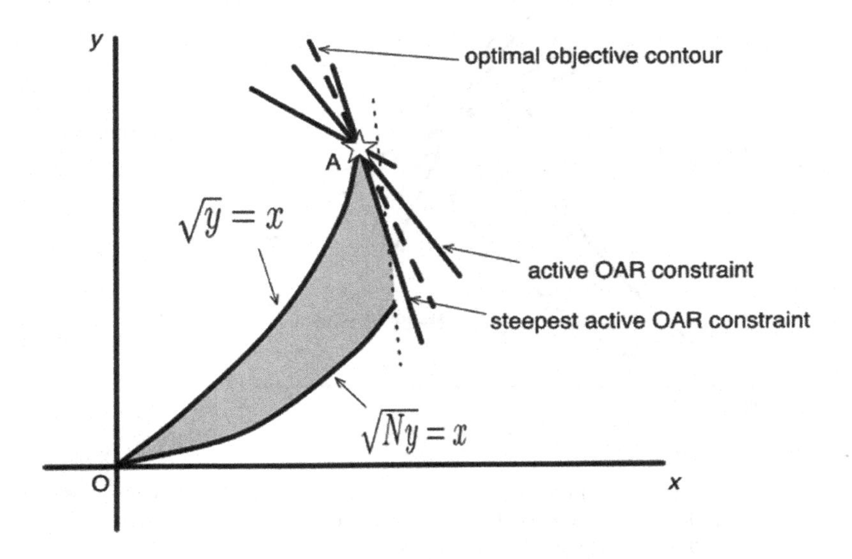

Figure 5.1 All active linear constraints at the optimal point $x^\star, y^\star$ (shown as a star labeled A) are displayed as solid lines. Inactive linear constraints are shown as thin dotted lines. The magnitude α_0/β_0 of the slope of the optimal objective contour (dashed line) must be no larger than the magnitude of the slope of the steepest active linear constraint. Adapted from [60, figure 2] per reuse policy available at https://publishingsupport.iopscience.iop.org/reusing-iop-published-content/.

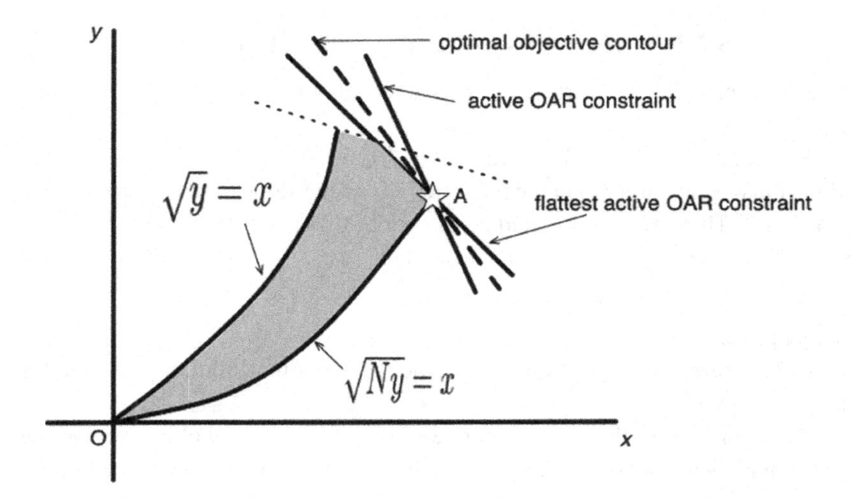

Figure 5.2 All active linear constraints at the optimal point $x^\star, y^\star$ (shown as a star labeled A) are displayed as solid lines. Inactive linear constraints are shown as thin dotted lines. The magnitude α_0/β_0 of the slope of the optimal objective contour (dashed line) must be no smaller than the magnitude of the slope of the flattest active linear constraint. Adapted from [60, figure 3] per reuse policy available at https://publishingsupport.iopscience.iop.org/reusing-iop-published-content/.

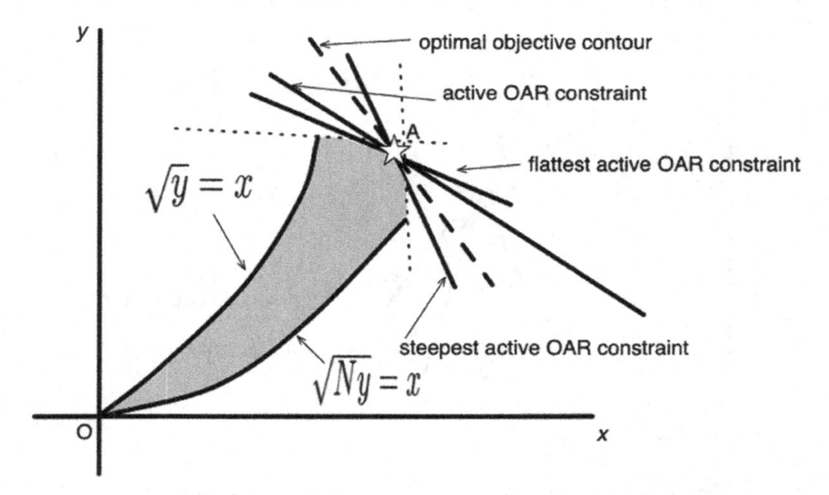

Figure 5.3 All active linear constraints at the optimal point $x^\star, y^\star$ (shown as a star labeled A) are displayed as solid lines. Inactive linear constraints are shown as thin dotted lines. The magnitude α_0/β_0 of the slope of the optimal objective contour (dashed line) must be between the magnitudes of the slopes of the flattest and steepest active linear constraints. Adapted from [60, figure 4] per reuse policy available at https://publishingsupport.iopscience.iop.org/reusing-iop-published-content/.

Table 5.1. *Data for a head-and-neck cancer example.*

m	OAR	T_m	δ_m (Gy)	s_m
1	Spinal cord	35	1.2857	0.65
2	Brainstem	35	1.4286	0.6
3	Left parotid	35	0.8	0.4
4	Right parotid	35	0.75	0.35
5	Unspecified normal tissue	35	2	1

Table 5.2. *Classification of five OAR based on which among conditions* (5.1)–(5.6) *they satisfy, for the head-and-neck cancer example.*

m	$s_m x^\star$ (Gy)	$T_m \delta_m$ (Gy)	$s_m^2 y^\star$ (Gy^2)	$T_m \delta_m^2$ (Gy^2)	cond. sat.	set
1	46.8	45	52.4745	57.8558	(5.6)	$\mathcal{M}_6$
2	43.2	50	44.7120	71.4314	(5.4)	$\mathcal{M}_4$
3	28.8	28	19.8720	22.4000	(5.6)	$\mathcal{M}_6$
4	25.2	26.25	15.2145	19.6875	(5.4)	$\mathcal{M}_4$
5	72	70	124.2	140	(5.6)	$\mathcal{M}_6$

The figure shows that the magnitude α_0/β_0 of the optimal objective contour slope must be in between the magnitudes of the slopes of the flattest and steepest active constraints. Thus, any $\alpha_0 > 0$ and $\beta_0 > 0$ values with

$$L \leq \frac{\alpha_0}{\beta_0} \leq U \tag{5.19}$$

are suitable.

The entire procedure is summarized in Algorithm 3. We apply this algorithm to an illustrative example in the next section.

5.3 Numerical Example

Consider an example with five OAR and data as in Table 5.1. We pursue the following question: what values of $\alpha_0 > 0$, $\beta_0 >$, and ρ_m, for $m \in \mathcal{M} = \{1,2,3,4,5\}$ would make a dosing plan with 1.8 Gy each in 30 sessions and 1.5 Gy each in 12 sessions optimal? We thus have $N = 30 + 12 = 42$, $x^\star = 30 \times 1.8 + 12 \times 1.5 = 72$ Gy, and $y^\star = 30 \times (1.8)^2 + 12 \times (1.5)^2 = 124.2$ Gy^2. Table 5.2 lists the resulting calculated values of $s_m x^\star$, $T_m \delta_m$, $s_m^2 y^\star$, and $T_m \delta_m^2$, for $m \in \{1,2,3,4,5\}$. The table also categorizes the five OAR into mutually exclusive sets $\mathcal{M}_1, \ldots, \mathcal{M}_6$ using conditions (5.1)–(5.6). The last

Algorithm 3 Exact closed-form solution of inverse problem

```
1: Input - number of sessions $N \geq 1$ and a nonnegative $\vec{d}^\star = (d_1^\star, \ldots, d_N^\star) \neq 0$.
2: Input - $T_m, \delta_m, s_m$ values for each $m \in \mathcal{M}$.
3: Let $x^\star = \sum_{t=1}^{N} d_t^\star$ and $y^\star = \sum_{t=1}^{N} (d_t^\star)^2$.
4: Verify that Assumptions 5.2 and 5.4 hold. If they don't, terminate execution.
5: Fix $\rho_m > 0$ arbitrarily, for $m \in \mathcal{M}_1$.
6: Interval for $\rho_m$: $(0, \infty)$, for $m \in \mathcal{M}_2 \cup \mathcal{M}_3 \cup \mathcal{M}_4$.
7: Let $\mu_m = \frac{-1}{\rho_m s_m}$, for $m \in \mathcal{M}_1$, as the slope of constraint (3.30).
8: Let $\underline{\mu} = \min_{m \in \mathcal{M}_1} -\mu_m$.
       ▷ Flattest slope magnitude; interpret $\infty$ if $\mathcal{M}_1 = \emptyset$.
9: Let $\bar{\mu} = \max_{m \in \mathcal{M}_1} -\mu_m$.
       ▷ Steepest slope magnitude; interpret $-\infty$ if $\mathcal{M}_1 = \emptyset$.
10: for each subset $\mathcal{M}_5^a$ of $\mathcal{M}_5$ do   ▷ (3.30) made active for all $m \in \mathcal{M}_5^a$.
11:   for each subset $\mathcal{M}_6^a$ of $\mathcal{M}_6$ do   ▷ (3.30) made active for all $m \in \mathcal{M}_6^a$.
12:     if $\mathcal{M}_1 \cup \mathcal{M}_5^a \cup \mathcal{M}_6^a \neq \emptyset$ then   ▷ Will be met per Assumption 5.4.
13:       Interval for $\rho_m$: $\left(0, \frac{T_m\delta_m - s_m x^\star}{s_m^2 y^\star - T_m \delta_m^2}\right)$, for $m \in \mathcal{M}_5 \setminus \mathcal{M}_5^a$.
              ▷ (3.30) inactive at $(x^\star, y^\star)$.
14:       Interval for $\rho_m$: $\left(\frac{T_m\delta_m - s_m x^\star}{s_m^2 y^\star - T_m \delta_m^2}, \infty\right)$, for $m \in \mathcal{M}_6 \setminus \mathcal{M}_6^a$.
              ▷ (3.30) inactive at $(x^\star, y^\star)$.
15:       Let $\mathcal{M}^a = \mathcal{M}_5^a \cup \mathcal{M}_6^a$.
16:       $\rho_m$: $\frac{T_m\delta_m - s_m x^\star}{s_m^2 y^\star - T_m \delta_m^2}$, for $m \in \mathcal{M}^a$.   ▷ (3.30) active at $(x^\star, y^\star)$.
17:       Let $\sigma_m = \frac{-1}{s_m}\left(\frac{s_m^2 y^\star - T_m \delta_m^2}{T_m\delta_m - s_m x^\star}\right)$, for $m \in \mathcal{M}^a$ (slope of (3.30)).
18:       Let $\underline{\sigma} = \min_{m \in \mathcal{M}^a} -\sigma_m$.
              ▷ Flattest slope magnitude; interpret $\infty$ if $\mathcal{M}^a = \emptyset$.
19:       Let $\bar{\sigma} = \max_{m \in \mathcal{M}^a} -\sigma_m$.
              ▷ Steepest slope magnitude; interpret $-\infty$ if $\mathcal{M}^a = \emptyset$.
20:       Let $L = \min\{\underline{\mu}, \underline{\sigma}\}$.
              ▷ Flattest slope magnitude over $\mathcal{M}_1 \cup \mathcal{M}^a$.
21:       Let $U = \max\{\bar{\mu}, \bar{\sigma}\}$.
              ▷ Steepest slope magnitude over $\mathcal{M}_1 \cup \mathcal{M}^a$.
22:       if $x^\star = \sqrt{y^\star}$ then   ▷ Claim 1, Lemma 2.2.
23:         Any $\alpha_0 > 0$, $\beta_0 > 0$ values with $\frac{\alpha_0}{\beta_0} \leq U$ are suitable.
24:       else if $x^\star = \sqrt{N y^\star}$ then   ▷ Claim 2, Lemma 2.2.
25:         Any $\alpha_0 > 0$, $\beta_0 > 0$ values with $\frac{\alpha_0}{\beta_0} \geq L$ are suitable.
26:       else   ▷ $\sqrt{y^\star} < x^\star < \sqrt{N y^\star}$ must hold (Claim 3, Lemma 2.2).
27:         Any $\alpha_0 > 0$, $\beta_0 > 0$ values with $L \leq \frac{\alpha_0}{\beta_0} \leq U$ are suitable.
28:       end if
29:     end if
30:   end for
31: end for
```

Table 5.3. *Each row lists parameter values that render $(x^\star, y^\star)$ optimal with a different combination of active constraints for the head-and-neck cancer example.*

$\mathcal{M}^a = \mathcal{M}_6^a$	ρ_1 (Gy^{-1})	ρ_2 (Gy^{-1})	ρ_3 (Gy^{-1})	ρ_4 (Gy^{-1})	ρ_5 (Gy^{-1})	$\frac{\alpha_0}{\beta_0}$ (Gy)
$\{1\}$	0.3346	$(0, \infty)$	$(0.3164, \infty)$	$(0, \infty)$	$(0.1266, \infty)$	4.5982
$\{3\}$	$(0.3346, \infty)$	$(0, \infty)$	0.3164	$(0, \infty)$	$(0.1266, \infty)$	7.9000
$\{5\}$	$(0.3346, \infty)$	$(0, \infty)$	$(0.3164, \infty)$	$(0, \infty)$	0.1266	7.9000
$\{1,3\}$	0.3346	$(0, \infty)$	0.3164	$(0, \infty)$	$(0.1266, \infty)$	$[4.5982, 7.9000]$
$\{1,5\}$	0.3346	$(0, \infty)$	$(0.3164, \infty)$	$(0, \infty)$	0.1266	$[4.5982, 7.9000]$
$\{3,5\}$	$(0.3346, \infty)$	$(0, \infty)$	0.3164	$(0, \infty)$	0.1266	$[4.5982, 7.9000]$
$\{1,3,5\}$	0.3344	$(0, \infty)$	0.3164	$(0, \infty)$	0.1266	$[4.5982, 7.9000]$

two columns of the table demonstrate that Assumption 5.2 holds. Lemma 5.1 then implies that there exist $\rho_m > 0$ values, for $m \in \{1, 2, 3, 4, 5\}$, that make $(x^\star, y^\star)$ feasible to (2VAR(N)).

The last two columns of Table 5.2 also show that $\mathcal{M}_1 = \mathcal{M}_2 = \mathcal{M}_3 = \mathcal{M}_5 = \emptyset$, $\mathcal{M}_4 = \{2, 4\}$, and $\mathcal{M}_6 = \{1, 3, 5\}$. Since $\mathcal{M}_6$ is not empty, Assumption 5.4 holds. All possible nonempty subsets $\mathcal{M}^a = \mathcal{M}_6^a$ of active constraints from the set $\{1, 3, 5\}$ and the corresponding positive parameter values that would render $(x^\star, y^\star)$ optimal to (2VAR(N)) are displayed in Table 5.3. The interval of possible ρ_2 and ρ_4 is $(0, \infty)$ because these two OAR belong to $\mathcal{M}_4$. Values of ρ_1, ρ_3, ρ_5 were obtained from Equation (5.13) for active constraints and Equation (5.12) for inactive constraints. Since the given plan is an unequal-dosage one, it yields $\sqrt{y^\star} = \sqrt{124.2} = 11.1445 < x^\star = 72 = \sqrt{5184} < \sqrt{5216.4} = \sqrt{42 \times 124.2} = \sqrt{Ny^\star}$, as expected. Values of α_0/β_0 in the final column of the table were therefore obtained via Equation (5.19) from Case 3.

Bibliographic Notes

A recent survey of inverse optimization is available in [32]. The question of which parameter values would make a particular dosing plan effective is mentioned, directly or indirectly, for example, in [29, 36, 39, 49, 95, 116, 126, 127, 131] and references therein. This chapter is based on [60]. That is currently the only inverse optimization approach for imputing radiobiological parameters from a dosing plan in a fractionation problem. A brief overview of this approach, without proofs and without a numerical example, was included in a tutorial [61]. That tutorial also included Algorithm 3. The numerical example is a modification of one of the three examples discussed in [60].

Exercises

Exercise 5.5 Sections 4.1 and 4.2 of Ghate [60] include inverse fractionation examples wherein the given dosing plans are equal dosage and single dosage, respectively. Solve those examples using Algorithm 3.

Exercise 5.6 Suppose estimates $\hat{\rho}_m$, for $m \in \mathcal{M}$, of the OAR dose-response parameters are available to the treatment planner. Similarly, an estimate $\hat{r}$ of the α_0/β_0 ratio for the tumor is also available. The treatment planner now wishes to minimize the nonnegative weighted sum of absolute value deviations $w_0|r-\hat{r}|+\sum_{m\in\mathcal{M}} w_m|\rho_m-\hat{\rho}_m|$ over all parameters that make a given dosing plan optimal. Describe whether or not and how the methodology from this chapter can be modified to solve this problem.

6

Spatiotemporally Integrated Fractionation

We employed tumor doses $d_1, d_2, \ldots, d_N$ in sessions $t = 1, 2, \ldots, N$ as decision variables thus far. When technology such as IMRT is utilized, this essentially implies that intensity profiles that administer these doses to the tumor and doses $s_m d_1, s_m d_2, \ldots, s_m d_N$ to each OAR $m \in \mathcal{M}$ are somehow available to the treatment planner. However, the problem of recovering such intensity profiles is very difficult and, in fact, may not even have a solution. One common approach in IMRT is to employ a linear dose-deposition model, whereby, as the name suggests, doses are linear functions of intensity profiles. Under this linear dose-deposition model, it is possible to recover intensity profiles that administer the aforementioned requisite doses to the tumor and OAR, by scaling any intensity profile that is available to the treatment planner. This indirect approach of recovering intensity profiles is often viewed as a spatiotemporally separated method. This is because the initial intensity profile is typically obtained without any temporal considerations – it is often derived only based on the anatomy, that is, the spatial characteristics of the cancerous region. Such spatiotemporally separated fractionation can lead to a lower biological effect on the tumor as compared to spatiotemporally integrated approaches. One spatiotemporally integrated approach to fractionation is described in this chapter.

6.1 Intractable Problem Formulation

In the linear dose-deposition model for IMRT, the tumor is (approximately) segmented into small cubes called voxels and the radiation field (or beam) in each treatment session is segmented into two-dimensional squares termed beamlets. Suppose that the tumor is composed of n voxels and the radiation field includes k beamlets. Let A denote an $n \times k$ nonnegative tumor dose-deposition matrix. Let $A_{i\cdot}$ denote its k-dimensional ith row, which corresponds to the ith voxel in the tumor, for $i = 1: n$. If intensity profile $u(t) \in \mathbb{R}^k_+$ is employed in the tth treatment session, then the ith voxel in the tumor receives a dose of $A_{i\cdot}u(t)$, for $i = 1: n$. Here, $\mathbb{R}^k_+$ is the set of nonnegative k-dimensional vectors. The nonnegative ℓth component $u_\ell(t)$ of $u(t) \in \mathbb{R}^k_+$ equals the radiation intensity of the ℓth beamlet, for $\ell = 1, \ldots, k$. Similarly, suppose OAR m includes n_m voxels, for $m \in \mathcal{M}$. Let A^m denote the $n_m \times k$ nonnegative dose-deposition matrix for this OAR and $A^m_{j\cdot}$ denote its jth row. This row corresponds to the jth voxel in OAR m. Thus, this voxel receives a dose of $A^m_{j\cdot}u(t)$ in the tth treatment session.

In this chapter, we will consider two types of OAR. The first are called serial OAR and the second are called parallel OAR. We denote the set of serial OAR by $\mathcal{M}_1 \subseteq \mathcal{M}$. As the name suggests, excessive toxic effect on even a small portion of a serial OAR hampers the biological function of that OAR. The spinal cord is one example of such an OAR. Thus, we put an upper bound on the BED administered to every voxel of a serial OAR. A parallel OAR, on the other hand, can retain its biological function as long as the proportion of its volume that suffers excessive toxic effects is not too large. Lungs are an example of a parallel OAR. This chapter accommodates two different ways to formulate BED constraints on such parallel OAR. The first is called the mean-BED constraint and the second is termed the dose-volume-BED constraint. The set of parallel OAR with the first type of constraints is denoted $\mathcal{M}_2 \subseteq \mathcal{M}$ and that with the second type is denoted $\mathcal{M}_3 \subseteq \mathcal{M}$. We assume for simplicity that sets $\mathcal{M}_1$, $\mathcal{M}_2$, and $\mathcal{M}_3$ are mutually exclusive.

Combining this notation with that developed in Chapters 2 and 3, we now formulate the spatiotemporally integrated fractionation problem as

(SP)

$$\min_{\substack{N \\ u(1),\ldots,u(N)}} \sum_{i=1}^{n} \exp\left(-\alpha_0 \sum_{t=1}^{N} (A_{i\cdot}u(t)) - \beta_0 \sum_{t=1}^{N} (A_{i\cdot}u(t))^2 + \tau(N) \right) \quad (6.1)$$

$$\sum_{t=1}^{N} \left[(A^m_{j\cdot}u(t)) + \rho_m (A^m_{j\cdot}u(t))^2 \right] \leq C_m,\ j = 1: n_m,\ m \in \mathcal{M}_1 \quad (6.2)$$

$$\sum_{t=1}^{N} \frac{\sum_{j=1}^{n_m} (A^m_{j\cdot}u(t)) + \rho_m (A^m_{j\cdot}u(t))^2}{n_m} \leq C_m, m \in \mathcal{M}_2 \quad (6.3)$$

$$\sum_{j=1}^{n_m} \mathbb{1}^m_j(u_1, \ldots, u_N) \leq \lfloor \phi_m n_m \rfloor,\ m \in \mathcal{M}_3 \quad (6.4)$$

$$S_\epsilon u(t) \leq 0,\ t = 1: N \quad (6.5)$$

$$u(t) \geq 0,\ t = 1\colon N \tag{6.6}$$

$$u(t) \in \mathbb{R}^k,\ t = 1\colon N \tag{6.7}$$

$$1 \leq N \leq N_{\max},\ \text{integer}. \tag{6.8}$$

Here, we have used the notation $C_m = T_m\delta_m(1+\rho_m\delta_m)$ that was introduced in Chapter 3. The ith term inside the sum in the objective function of formulation (SP) equals the number of tumor cells that survive within the ith voxel in the tumor. The objective function thus calculates the total number of tumor cells that remain (TNTCR) at the end of the treatment course. We have assumed here, for notational simplicity, that each tumor voxel includes an identical number of tumor cells at the beginning of the treatment course. This assumption can be relaxed simply by multiplying the ith term inside the sum with an initial number of cells that depends on i, if such an estimate is available.

Constraint (6.2) ensures that the BED of radiation intensity vectors $u_1, \ldots, u_N$ for each voxel j in serial OAR $m \in \mathcal{M}_1$ is no more than the tolerable limit of C_m.

Constraint (6.3) ensures that the average BED of radiation intensity vectors $u_1, \ldots, u_N$ over all voxels within parallel OAR $m \in \mathcal{M}_2$ is no more than the tolerable limit of C_m.

The formulation uses binary-valued functions

$$\mathbb{1}_j^m(u_1, \ldots, u_N) = \begin{cases} 1, \text{ if } \sum_{t=1}^{N}\left[(A_{j\cdot}^m u(t)) + \rho_m (A_{j\cdot}^m u(t))^2\right] > C_m, \\ 0, \text{ otherwise,} \end{cases} \tag{6.9}$$

for $j = 1\colon n_m$ and $m \in \mathcal{M}_3$. That is, the function value is 1 if the BED of radiation intensity vectors $u_1, \ldots, u_N$ for voxel j in parallel OAR $m \in \mathcal{M}_3$ exceeds the tolerable limit of C_m. The expression $\sum_{j=1}^{n_m} \mathbb{1}_j^m(u_1, \ldots, u_N)$ on the left-hand side of constraint (6.4) thus counts the number of voxels in OAR $m \in \mathcal{M}_3$ whose BED exceeds the tolerable limit. Constraint (6.4) thus captures the dose-volume protocol informally mentioned above that no more than a fraction ϕ_m of the voxels within OAR $m \in \mathcal{M}_3$ should receive a BED above the tolerable limit.

The formulation includes smoothness constraints on radiation intensity vectors $u(1), \ldots, u(N)$, to ensure that they are deliverable in practice using available technology. Specifically, we put an upper bound of $0 < \epsilon \leq 1$ on the absolute relative difference between intensities of each pair of nearest neighbor beamlets. Thus, larger values of ϵ allow for bigger differences between the intensities of neighboring beamlets. This in turn allows the treatment planner

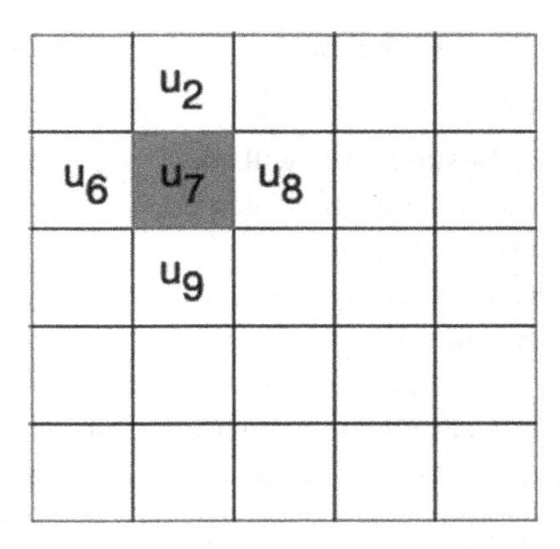

Figure 6.1 Schematic of a radiation field $u \in \mathbb{R}^{25}$ with 25 beamlets. A session-label such as t is not included for brevity since it is not relevant here. Beamlets $2, 6, 8, 9$ are nearest neighbors of the beamlet numbered 7 that is shaded in gray. Formulation (SP) includes smoothness constraints of the form $|u_2 - u_7| \leq \epsilon u_7$ on the radiation intensities of each of these nearest neighbors. These constraints can be rewritten as the pair $u_2 - (1+\epsilon)u_7 \leq 0$ and $(1-\epsilon)u_7 - u_2 \leq 0$ of linear constraints. Matrix S_ϵ includes the coefficients of all beamlet intensities in such pairs of smoothness constraints.

to administer dosing plans that conform better to the shape of the tumor. This achieves higher tumor damage and better OAR protection. The nearest neighbor relationships among beamlets are provided as input in the form of a $k \times k$ binary adjacency matrix, where the entry (i, j) in row i and column j is 1 if beamlet j is a nearest neighbor of beamlet i; the entry is 0 otherwise. Such an adjacency matrix has a block diagonal structure, where each block corresponds to one beam. See Figure 6.1, which explains that the smoothness constraints can be expressed in linear form as $S_\epsilon u(t) \leq 0$, where S_ϵ is called a smoothness matrix composed of entries $-(1+\epsilon), (1-\epsilon), -1, 0, +1$. This is captured in constraint (6.5).

Without loss of generality, we assume that for every OAR $m \in \mathcal{M}$, every row of its dose-deposition matrix A^m has at least one strictly positive entry. For if not, then the OAR voxel corresponding to a row with all zeros in A^m can be removed from further consideration. This is because a row of zeros implies that none of the beamlets deposit any dose in that voxel. We also make two technical assumptions.

Assumption 6.1 *The set $\mathcal{M}_1 \cup \mathcal{M}_2$ is not empty. That is, there is at least one serial OAR or a parallel OAR with a mean-BED constraint.*

Assumption 6.2 *There exists an OAR $m \in \mathcal{M}_1 \cup \mathcal{M}_2$ such that every column of its dose-deposition matrix A^m has at least one positive entry.*

These two technical assumptions will help us establish the existence of optimal solutions to the fractionation problem in Lemma 6.3. For if such an OAR does not exist, and if, for example, $\mathcal{M}_3 = \emptyset$, then, informally speaking, the intensity of the beamlet corresponding to a zero column can be increased arbitrarily without bound. It seems difficult to formally establish such unboundedness of a beamlet intensity when smoothness constraints are present, so we eliminate this pathological possibility via the above two technical assumptions. These two technical assumptions typically hold in practice simply because every beamlet of interest deposits at least some dose in some part of the patient's body (and hence its intensity should not be allowed to grow too high).

Lemma 6.3 *Problem (SP) has an optimal solution.*

Proof Since there are a finite number of possible values of N, it suffices to show that the problem has an optimal solution for each fixed N. Thus, consider any fixed integer $1 \leq N \leq N_{\max}$ and denote the resulting problem by (SP(N)). For every $m \in \mathcal{M}_3$, the BED for at least $L_m = n_m - \lfloor \phi_m n_m \rfloor$ and at most n_m voxels should be less than the tolerance limit. Thus, there are $W_m = \sum_{\ell=L_m}^{n_m} \binom{n_m}{\ell}$ distinct ways to express the dose-volume constraint (6.4), depending on which voxels are forced to receive a BED below the tolerance limit. We index these distinct ways as $w_m = 1, 2, \ldots, W_m$. Let $\mathcal{N}_m(w_m)$ denote the set of voxels for which the tolerance limit is enforced. Thus, there are $W = \prod_{m \in \mathcal{M}_3} W_m$ ways to express the dose-volume constraints. Let $\vec{w} = (w_1, \ldots, w_{|\mathcal{M}_3|})$. We thus create W subproblems, each corresponding to one of these W distinct ways of expressing the dose-volume constraints. We write these subproblems as

(SP($\vec{w}, N$))

$$\min_{u(1),\ldots,u(N)} \sum_{i=1}^{n} \exp\left(-\alpha_0 \sum_{t=1}^{N} (A_{i\cdot} u(t)) - \beta_0 \sum_{t=1}^{N} (A_{i\cdot} u(t))^2 + \tau(N) \right)$$

(6.2) – (6.3)

$$\sum_{t=1}^{N} \left[(A^m_{j\cdot} u(t)) + \rho_m (A^m_{j\cdot} u(t))^2 \right] \leq C_m, \; j \in \mathcal{N}_m(w_m), \; m \in \mathcal{M}_3$$

(6.5) – (6.7).

An optimal solution to (SP(N)) can be recovered by comparing the objective values of optimal solutions to these W subproblems. Thus, it suffices to show that (SP($\vec{w}, N$)) has an optimal solution, for each $\vec{w}$. Fix $\vec{w}$. Let $U(\vec{w}, N) \in \mathbb{R}^{kN}$ denote the set of feasible intensity vectors $(u(1), \ldots, u(N))$ to this problem. This set is nonempty because radiation fields wherein every beamlet intensity is zero in every treatment session are feasible. In light of Assumptions 6.1 and 6.2, suppose a serial OAR $m \in \mathcal{M}_1$ has a positive entry in each column of its dose-deposition matrix A^m (if instead a parallel OAR with mean-BED constraints possesses this property, the proof can be adapted easily). For each beamlet $\ell = 1, \ldots, k$, let J_ℓ denote the set of rows of A^m that include a positive entry in column ℓ. For each $i \in J_\ell$, we denote the corresponding positive entry in column ℓ by $A^n_{i\ell}$. Since every entry in A^m is nonnegative, the maximum dose constraint (6.2) implies that $u_\ell(t) \leq \frac{T_m \delta_m (1+\rho_m \delta_m)}{\min\limits_{i \in J_\ell} A^m_{i\ell}}$, for each beamlet ℓ in each session t. Thus, the feasible region is bounded in $\mathbb{R}^{kN}_+$. It is also closed since all constraint functions are continuous in $(u(1), \ldots, u(N))$ over $\mathbb{R}^{kN}$. Similarly, the objective function is continuous. Problem (SP($\vec{w}, N$)) thus has an optimal solution, and the proof is complete. □

Unfortunately, formulation (SP) is computationally intractable for several reasons. First, it includes an integer variable N and continuous variables $u(1), \ldots, u(N)$. Thus, it is natural to first solve this problem for each fixed value of N and then compare objective values across different N. However, even when N is fixed in this manner, the problem remains computationally challenging, partly because the number of decision variables in vectors $u(1), \ldots, u(N)$, which equals $k \times N$, is very large. To get a sense of the large scale of this problem, recall from Chapter 1 that radiation beams from multiple directions are often employed in IMRT. For example, IMRT for head-and-neck cancer may include beams from as many as seven directions. Each beam may be segmented into several thousand beamlets. Consequently, the order of magnitude of k, the total number of beamlets in each treatment session, is in the ten thousands. Common treatment protocols for head-and-neck cancer include over $N = 30$ treatment sessions. This means that the decision variables number in the several hundred thousands for such fixed values of N. Two other features of the formulation render it computationally difficult, even when N is fixed. Firstly, the TNTCR objective function is in general nonconvex, although constraints (6.2) and (6.3) are convex quadratic in $(u(1), \ldots, u(N))$. Moreover, dose-volume constraints (6.4) are not amenable to nonlinear programming algorithms, as they utilize somewhat complicated binary-valued functions as described in (6.9). We therefore describe an alternative simpler formulation of the problem in the next section.

6.2 Alternative Formulation and Approximate Solution Algorithm

The alternative formulation is obtained by first assuming that the treatment planner employs identical intensity profiles across different treatment sessions. This is consistent with current practice. We thus substitute $u(1) = u(2) = \cdots = u(N) = u$, for some $u \in \mathbb{R}^k$, in (SP) to yield

$$\min_{N,u} \sum_{i=1}^{n} \exp(-\alpha_0 N(A_{i}.u) - \beta_0 N(A_{i}.u)^2 + \tau(N)) \tag{6.10}$$

$$N(A^m_{j}.u) + N\rho_m(A^m_{j}.u)^2 \leq C_m, \; j = 1 \colon n_m; \; m \in \mathcal{M}_1 \tag{6.11}$$

$$N\frac{\left[\sum_{j=1}^{n_m}(A^m_{j}.u) + \rho_m(A^m_{j}.u)^2\right]}{n_m} \leq C_m, \; m \in \mathcal{M}_2 \tag{6.12}$$

$$\sum_{j=1}^{n_m} \mathbb{1}^m_j(N,u) \leq \lfloor \phi_m n_m \rfloor, \; m \in \mathcal{M}_3 \tag{6.13}$$

$$S_\epsilon u \leq 0 \tag{6.14}$$

$$u \geq 0 \tag{6.15}$$

$$u \in \mathbb{R}^k \tag{6.16}$$

$$1 \leq N \leq N_{\max}, \text{ integer.} \tag{6.17}$$

The binary-valued function $\mathbb{1}^m_j(N,u)$ is obtained from $\mathbb{1}^m_j(u_1,\ldots,u_N)$ by substituting $u(1) = u(2) = \cdots = u(N) = u$ on the right-hand side of (6.9).

Now, in order to facilitate solution of this problem, we approximate the objective function with the TNTCR value achieved when an identical dose is administered to each tumor voxel. We use the average tumor dose $\sum_{i=1}^{n} \frac{A_{i}.u}{n} = \bar{A}u$ over all voxels as this uniform dose. Here, $\bar{A} = \sum_{i=1}^{n} \frac{A_{i}.}{n}$ is a k-dimensional vector whose component $\bar{A}_\ell$ is formed by averaging entries in the ℓth column of matrix A. This changes the objective function to

$$n \exp(-\alpha_0 N(\bar{A}u) - \beta_0 N(\bar{A}u)^2 + \tau(N)). \tag{6.18}$$

Minimizing this objective is equivalent to maximizing

$$\alpha_0 N(\bar{A}u) + \beta_0 N(\bar{A}u)^2 - \tau(N). \tag{6.19}$$

Moreover, when N is fixed, this is equivalent to maximizing $\alpha_0 N(\bar{A}u) + \beta_0 N(\bar{A}u)^2$. Since this quantity is monotonic in $\bar{A}u$, it suffices to simply maximize this average tumor dose. Thus, we now focus on solving the problem

$$(\tilde{\text{SP}}(N)) \max_u \bar{A}u \tag{6.20}$$

$$N(A^m_{j.}u) + N\rho_m (A^m_{j.}u)^2 \leq C_m, \; j = 1: n_m; \; m \in \mathcal{M}_1 \tag{6.21}$$

$$N\frac{\left[\sum_{j=1}^{n_m}(A^m_{j.}u) + \rho_m (A^m_{j.}u)^2\right]}{n_m} \leq C_m, \; m \in \mathcal{M}_2 \tag{6.22}$$

$$\sum_{j=1}^{n_m} \mathbb{1}^m_j(N,u) \leq \lfloor \phi_m n_m \rfloor, \; m \in \mathcal{M}_3 \tag{6.23}$$

$$(6.14) - (6.16). \tag{6.24}$$

This problem has an optimal solution by logic similar to that in the proof of Lemma 6.3. The objective function in this problem is linear in u. Since $A^m_{j.}u$ is nonnegative and the left-hand side of the serial OAR constraint $N(A^m_{j.}u) + N\rho_m(A^m_{j.}u)^2 \leq C_m$ is monotone in $A^m_{j.}u$, this quadratic constraint is satisfied if, and only if, the linear constraint $A^m_j u \leq \frac{-1+\sqrt{1+4\rho_m C_m/N}}{2\rho_m}$ holds. Recall that the mean-BED constraint on the parallel OAR in $\mathcal{M}_2$ is convex quadratic. Thus, the dose-volume constraints (6.23) are the main remaining challenge in solving this problem.

We implement a simple approach to circumvent this hurdle and obtain an approximate solution. We first solve $(\tilde{\text{SP}}(N))$ without the dose-volume constraints (6.23). That is, we solve

$$(\tilde{\text{SP}}1(N)) \max_u \bar{A}u$$

$$A^m_{j.}u \leq \frac{-1+\sqrt{1+4\rho_m C_m/N}}{2\rho_m}, \; j = 1: n_m; \; m \in \mathcal{M}_1$$

$$\sum_{j=1}^{n_m}(A^m_{j.}u) + \rho_m(A^m_{j.}u)^2 \leq n_m C_m/N, \; m \in \mathcal{M}_2$$

$$(6.14) - (6.16).$$

Note that this problem is convex. It has an optimal solution by logic similar to that in the proof of Lemma 6.3. We use u^N to denote its optimal solution. We then calculate the dose $A^m_j u^N$ administered to each voxel j in each parallel OAR m in $\mathcal{M}_3$. We then find $n_m - \lfloor \phi_m n_m \rfloor$ voxels within each OAR m in $\mathcal{M}_3$ that receive the lowest doses. We use $\mathcal{N}_m$ to denote the set of these voxels. Then, we solve the optimization problem again by including BED upper limit constraints on all voxels from these sets, as this ensures that the dose-volume-BED constraints (6.23) are satisfied. These constraints are of the form $(A^m_{j\cdot}u) + \rho_m(A^m_{j\cdot}u)^2 \leq C_m$, and, as explained earlier, can be rewritten as $A^m_{j\cdot}u \leq \frac{-1+\sqrt{1+4\rho_m C_m/N}}{2\rho_m}$. In particular, we solve the problem

$$(\tilde{\text{SP}}2(N)) \ \max_u \ \bar{A}u$$

$$A^m_{j\cdot}u \leq \frac{-1+\sqrt{1+4\rho_m C_m/N}}{2\rho_m}, \ j = 1: n_m; \ m \in \mathcal{M}_1$$

$$\sum_{j=1}^{n_m}(A^m_{j\cdot}u) + \rho_m(A^m_{j\cdot}u)^2 \leq n_m C_m/N, \ m \in \mathcal{M}_2$$

$$A^m_{j\cdot}u \leq \frac{-1+\sqrt{1+4\rho_m C_m/N}}{2\rho_m}, \ j \in \mathcal{N}_m; \ m \in \mathcal{M}_3$$

$$(6.14) - (6.16).$$

This problem is convex. It has an optimal solution by logic similar to that in the proof of Lemma 6.3. We again use u^N to represent its optimal solution, with some abuse of notation. We then evaluate the objective function value (6.19) in the alternative formulation using this solution. We repeat this procedure for all positive integers N between 1 and $N_{\max}$ and find an N that attains the largest objective value. We report this N and the corresponding u^N as an approximate solution for the alternative formulation (6.10)–(6.17). The overall procedure is summarized in Algorithm 4. We implement this procedure on a representative example in the next section.

6.3 Numerical Experiments

We consider a hypothetical, two-dimensional, toy example with an L-shaped tumor and a single square OAR as shown in Figure 6.2. This example includes $n = 12$ tumor voxels, $n_1 = 4$ OAR voxels, and a total of $k = 8$ beamlets. An 8×8 adjacency matrix of nearest neighbor relationships among the beamlets is displayed in Table 6.1. We model the single OAR as a serial one, so $\mathcal{M} = \mathcal{M}_1 = \{1\}$, $\mathcal{M}_2 = \emptyset$, and $\mathcal{M}_3 = \emptyset$.

The conventional number of treatment sessions, T_m, was set to 35 for the OAR. The tolerance dose per session for the spinal cord was set at $\delta_1 = 1.2857$ Gy. This corresponds to a total tolerance dose of 45 Gy, which is typical of spinal cord in head-and-neck cancer. The maximum possible number of treatment sessions was set at $N_{\max} = 100$. The tumor dose-response parameters were fixed at $\alpha_0 = 0.35$ Gy^{-1} and $\beta_0 = 0.035$ Gy^{-2}.

Algorithm 4 Approximate solution of (6.10)–(6.17)

1: Input - $n \times k$ tumor dose-deposition matrix A; tumor parameters $\alpha_0, \beta_0, T_{\text{double}}$; $N_{\max}$.
2: Input - set $\mathcal{M}$ of OAR, with mutually exclusive and exhaustive subsets $\mathcal{M}_1, \mathcal{M}_2, \mathcal{M}_3$.
3: Input - $n_m \times k$ OAR dose-deposition matrices A^m, for each $m \in \mathcal{M}$.
4: Input - T_m, δ_m, ρ_m, for each $m \in \mathcal{M}$.
5: Input - $0 \leq \phi_m \leq 1$, for each $m \in \mathcal{M}_3$.
6: Input - smoothness parameter $0 < \epsilon \leq 1$; beamlet adjacency matrix I.
7: Construct smoothness matrix S_ϵ.
8: best= $-\infty$.
9: **for** $N = 1 : N_{\max}$ **do**
10: Solve problem ($\tilde{\text{SP}}1(N)$) to obtain solution u^N.
11: **for** each OAR $m \in \mathcal{M}_3$ **do**
12: $\mathcal{N}_m$=set of $\lfloor \phi_m n_m \rfloor$ voxels j in OAR m with smallest doses $A^m_{j\cdot} u^N$.
13: **end for**
14: If $\mathcal{M}_3 \neq \emptyset$, solve problem ($\tilde{\text{SP}}2(N)$) to obtain solution u^N.
15: Calculate $f^\star(N) = N\alpha_0(\bar{A}u^N) + N\beta_0(\bar{A}u^N)^2 - \tau(N)$.
16: **if** $f^\star(N) >$best **then** ▷ A better objective value found.
17: best= $f^\star(N)$.
18: $N_{\text{optimal}} = N$.
19: $u^{\text{optimal}} = u^N$.
20: **end if**
21: **end for**
22: Output - $N_{\text{optimal}}, u^{\text{optimal}}$ as an approximate solution.

Table 6.2 displays the optimal number of treatment sessions and the corresponding optimal objective value attained by Algorithm 4, for different values of $T_{\text{double}}, \epsilon, \rho_1$. The table shows that when ϵ and ρ_1 are fixed, the optimal number of treatment sessions increases with T_{double}. This is consistent with the previous chapters, as larger values of T_{double} model slower tumor growth. The optimal objective value increases with increasing T_{double} values. This is also consistent with the previous chapters, as any feasible treatment plan produces a larger objective value when the tumor grows slower. When $T_{\text{double}}, \rho_1$ are fixed, the optimal objective increases with increasing values of smoothness parameter ϵ. Mathematically, this is because the feasible region of the problem is larger for higher values of the smoothness parameter. This is also intuitive because larger values of the smoothness parameter allow for increased modulation of intensity profiles. A reduction in the smoothness of

Table 6.1. *Block-diagonal adjacency matrix for Figure 6.2.*

			Beam 1				Beam 2			
			beamlets				beamlets			
			1	2	3	4	5	6	7	8
Beam1	beamlets	1	0	1	0	0	0	0	0	0
		2	1	0	1	0	0	0	0	0
		3	0	1	0	1	0	0	0	0
		4	0	0	1	0	0	0	0	0
Beam2	beamlets	5	0	0	0	0	0	1	0	0
		6	0	0	0	0	1	0	1	0
		7	0	0	0	0	0	1	0	1
		8	0	0	0	0	0	0	1	0

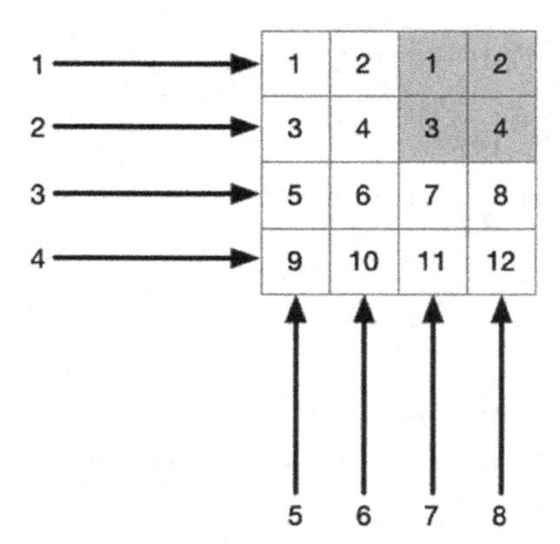

Figure 6.2 A hypothetical, two-dimensional, L-shaped tumor with a single square OAR shaded in gray. Tumor voxels are numbered 1–12 and OAR voxels are numbered 1–4. There are two beams with beamlets (black arrows) numbered 1–4 and 5–8, respectively. The adjacency structure for these eight beamlets is displayed in Table 6.1.

the optimal intensity profiles across $\epsilon = 0.1, 0.5, 1$ is illustrated in Figure 6.3. The optimal intensity values were normalized to belong to the interval $[0, 1]$ by dividing them with the maximum intensity value for each ϵ. This allows us to visually compare the level of intensity modulation across different values of ϵ. Figure 6.3 illustrates that the dose conforms much better to the tumor when the smoothness parameter is large and consequently the intensity profile is better modulated. The optimal number of treatment sessions appears to increase as the value of ρ_1 increases in Table 6.2. This trend is similar to Tables 2.1 and 2.2.

Table 6.2. *($N_{optimal}, f^{\star}(N_{optimal})$) for different $T_{double}, \epsilon, \rho_1$ values, when the single OAR is modeled as a serial one.*

		$\rho_1 = 1/5\ \text{Gy}^{-1}$					
		T_{double} (days)					
		2	3	4	6	8	10
ϵ	0.01	(5, 12.83)	(7, 13.39)	(10, 13.82)	(14, 14.45)	(18, 14.89)	(22, 15.22)
	0.05	(5, 13.53)	(7, 14.09)	(10, 14.52)	(14, 15.14)	(18, 15.58)	(22, 15.92)
	0.1	(5, 14.43)	(7, 14.98)	(10, 15.41)	(14, 16.03)	(18, 16.47)	(22, 16.80)
	0.5	(4, 22.34)	(6, 22.77)	(8, 23.11)	(12, 23.63)	(16, 24.00)	(19, 24.28)
	1	(1, 35.94)	(1, 35.94)	(1, 35.94)	(3, 36.00)	(4, 36.06)	(5, 36.11)
		$\rho_1 = 1/4\ \text{Gy}^{-1}$					
		T_{double} (days)					
		2	3	4	6	8	10
ϵ	0.01	(6, 11.99)	(9, 12.71)	(12, 13.27)	(18, 14.06)	(23, 14.63)	(28, 15.05)
	0.05	(6, 12.65)	(9, 13.38)	(12, 13.94)	(18, 14.75)	(24, 15.31)	(28, 15.74)
	0.1	(6, 13.49)	(9, 14.23)	(13, 14.80)	(18, 15.61)	(24, 16.19)	(29, 16.62)
	0.5	(6, 20.81)	(9, 21.55)	(13, 22.13)	(19, 22.95)	(24, 23.53)	(29, 23.97)
	1	(4, 32.59)	(7, 33.09)	(9, 33.49)	(14, 34.10)	(19, 34.54)	(23, 34.88)
		$\rho_1 = 1/3\ \text{Gy}^{-1}$					
		T_{double} (days)					
		2	3	4	6	8	10
ϵ	0.01	(7, 11.08)	(11, 11.99)	(15, 12.69)	(23, 13.72)	(30, 14.44)	(36, 14.99)
	0.05	(7, 11.70)	(12, 12.63)	(16, 13.35)	(23, 14.39)	(30, 15.13)	(37, 15.69)
	0.1	(7, 12.48)	(12, 13.45)	(16, 14.18)	(24, 15.25)	(31, 16.01)	(37, 16.58)
	0.5	(9, 19.31)	(13, 20.44)	(18, 21.28)	(27, 22.49)	(34, 23.33)	(41, 23.96)
	1	(9, 30.05)	(14, 31.18)	(18, 32.03)	(27, 33.24)	(34, 34.09)	(41, 34.72)

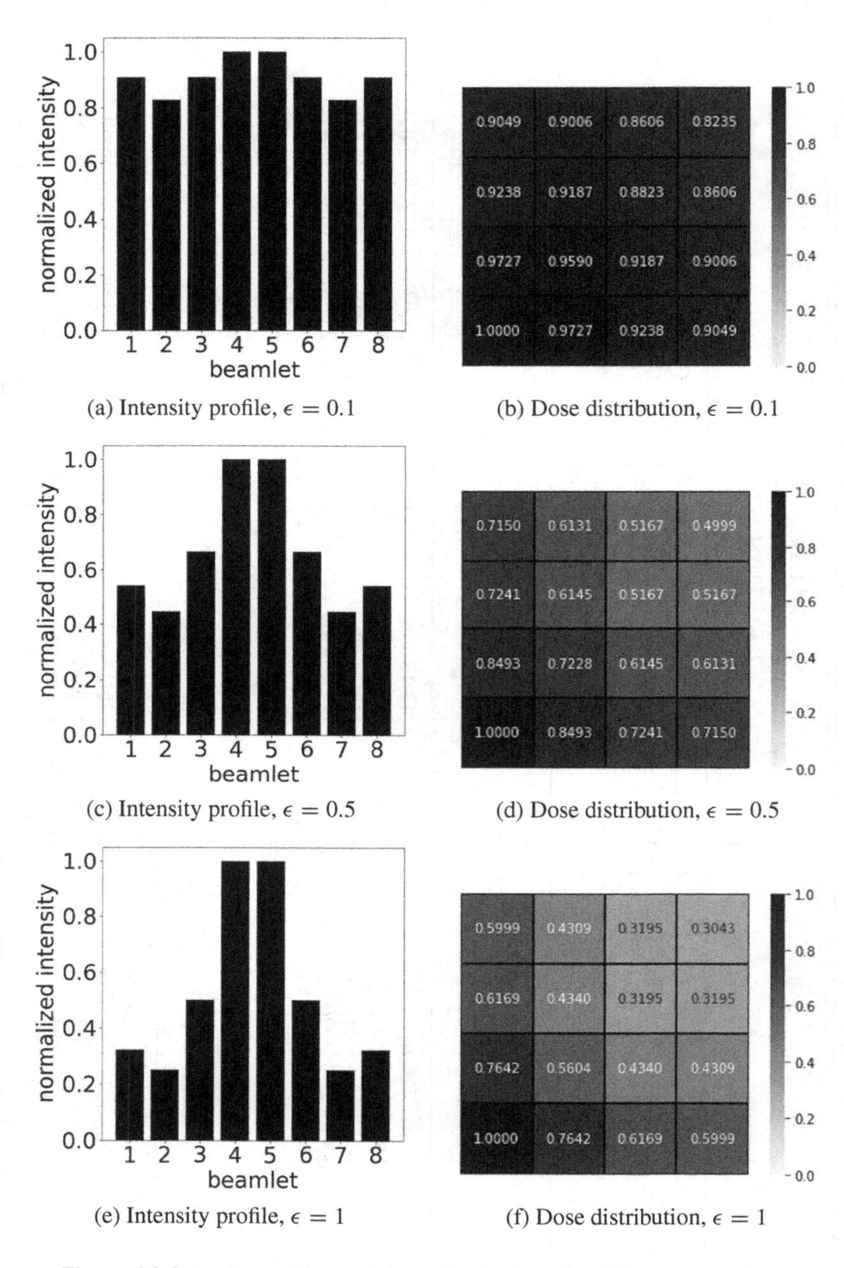

(a) Intensity profile, $\epsilon = 0.1$ (b) Dose distribution, $\epsilon = 0.1$

(c) Intensity profile, $\epsilon = 0.5$ (d) Dose distribution, $\epsilon = 0.5$

(e) Intensity profile, $\epsilon = 1$ (f) Dose distribution, $\epsilon = 1$

Figure 6.3 Intensity profiles and dose distributions for different smoothness parameters ϵ, for the case of $T_{\text{double}} = 10$ days and $\rho_1 = 1/5$ Gy^{-1}. Optimal intensity profile is better modulated and dose distribution is more conformal as ϵ increases. Symmetry in the L-shaped geometry is reflected throughout.

Table 6.3. *($N_{optimal}$, $f^{\star}(N_{optimal})$) for different T_{double}, ϵ, ρ_1 values, when the single OAR is modeled as a parallel one with a mean-BED constraint.*

		$\rho_1 = 1/5$ Gy^{-1}					
		T_{double} (days)					
		2	3	4	6	8	10
ϵ	0.01	(5, 14.10)	(7, 14.66)	(10, 15.09)	(14, 15.72)	(19, 16.16)	(22, 16.49)
	0.05	(5, 15.07)	(7, 15.63)	(10, 16.06)	(14, 16.69)	(18, 17.13)	(22, 17.46)
	0.1	(5, 16.36)	(7, 16.91)	(10, 17.34)	(14, 17.95)	(18, 18.39)	(22, 18.72)
	0.5	(2, 29.86)	(4, 30.09)	(6, 30.30)	(9, 30.64)	(11, 30.90)	(14, 31.10)
	1	(1, 57.36)	(1, 57.36)	(1, 57.36)	(1, 57.36)	(1, 57.36)	(1, 57.36)
		$\rho_1 = 1/4$ Gy^{-1}					
		T_{double} (days)					
		2	3	4	6	8	10
ϵ	0.01	(6, 13.18)	(9, 13.92)	(13, 14.49)	(18, 15.31)	(24, 15.89)	(29, 16.32)
	0.05	(6, 14.10)	(9, 14.85)	(13, 15.43)	(19, 16.26)	(24, 16.84)	(29, 17.28)
	0.1	(6, 15.30)	(10, 16.07)	(13, 16.65)	(19, 17.49)	(24, 18.08)	(29, 18.52)
	0.5	(6, 27.56)	(9, 28.25)	(12, 28.78)	(18, 29.55)	(23, 30.10)	(27, 30.51)
	1	(1, 50.90)	(1, 50.90)	(2, 50.93)	(3, 51.02)	(5, 51.11)	(6, 51.19)
		$\rho_1 = 1/3$ Gy^{-1}					
		T_{double} (days)					
		2	3	4	6	8	10
ϵ	0.01	(7, 12.20)	(12, 13.16)	(16, 13.89)	(24, 14.95)	(31, 15.70)	(37, 16.27)
	0.05	(8, 13.05)	(12, 14.04)	(16, 14.80)	(24, 15.89)	(31, 16.66)	(38, 17.24)
	0.1	(8, 14.19)	(12, 15.21)	(17, 15.99)	(25, 17.11)	(32, 17.90)	(39, 18.50)
	0.5	(9, 25.56)	(14, 26.76)	(19, 27.65)	(28, 28.93)	(36, 29.82)	(43, 30.49)
	1	(7, 45.73)	(12, 46.69)	(16, 47.44)	(24, 48.52)	(31, 49.29)	(38, 49.87)

We now repeat the above experiments with one change. We model the single OAR as a parallel one with a mean-BED constraint. That is, we now have $\mathcal{M}_1 = \emptyset$, $\mathcal{M}_2 = \{1\}$, and $\mathcal{M}_3 = \emptyset$. To bring forth the resulting additional flexibility afforded to the treatment planner as compared to the serial OAR above, we do not alter any other problem parameters. The results are displayed in Table 6.3. The qualitative trends in this table are the same as those in Table 6.2, as expected. The additional flexibility provided by the mean-BED constraint as compared to the max-BED constraint does indeed result in higher objective function values in Table 6.3 compared to Table 6.2. The added benefit of better modulation of the intensity profile seems to be higher in Table 6.3 than in Table 6.2. For example, when $\rho_1 = 1/4$ Gy^{-1} and $T_{\text{double}} = 2$ days in Table 6.2, the objective value increases from about 12 to about 33 when ϵ increases from 0.01 to 1. In Table 6.3, it increases from about 13 to about 50. A similar trend can be observed in other columns and rows of the two tables.

Bibliographic Notes

A description of linear dose-deposition models can be found in [34, 37, 129]. An algorithm that attempts to directly solve the formulation in Section 6.1, without simplifications, is not available in the literature. The simplified formulation and approximate solution method in Section 6.2 were developed in [105]. Extensive numerical experiments and sensitivity analyses with more realistic larger scale head-and-neck and prostate cases with dose-volume constraints were reported there. That paper used three-dimensional anatomies and two-dimensional radiation fields with smoothness matrices in all experiments. The L-shaped geometry and the corresponding dose-deposition matrices were constructed in [76] and utilized for numerical experiments in a different context in [76, 78]. A slightly different version of Figure 6.2 appeared in [76, 78]. Variants of the formulation in Section 6.2, including those that do not assume $u(1) = u(2) = \cdots = u(N)$, are discussed with corresponding approximate solution methods in [1, 2, 5, 56, 122].

There is an unrelated body of work on adapting intensity profiles in later sessions to compensate for errors made in earlier ones [21, 30, 31, 43, 87, 112, 133]. These errors may stem from uncertainty in organ motion within and across sessions and from variations in setting up the patient. Those papers did not utilize a dose-response model because their focus was not on radiobiological aspects of the problem. They are thus beyond the scope of this chapter.

Exercises

Exercise 6.4 Repeat the experiment reported in Tables 6.2 and 6.3 for the case where the single OAR is a parallel one with a dose-volume-BED constraint. Use $\phi_1 \in \{0, 1/4, 1/2, 3/4\}$ in your experiments. State whether or not you notice something special about the results for $\phi_1 = 0$. Discuss any qualitative trends in your results.

Exercise 6.5 The average tumor dose over all voxels was employed to simplify problem (6.10)–(6.17) in this chapter. Instead utilize the minimum tumor dose over all voxels to derive the corresponding counterparts of problems ($\tilde{\text{SP}}(N)$), ($\tilde{\text{SP}}1(N)$), ($\tilde{\text{SP}}2(N)$), and Algorithm 4. Then, repeat the numerical experiments in Section 6.3 with this setup.

7
Robust Spatiotemporally Integrated Fractionation

In this chapter, we present a robust optimization approach to incorporate uncertainty in the dose-response parameters of the spatiotemporally integrated formulation from Section 6.2. Thus, this chapter builds upon the previous one along the same lines as Chapter 4 built upon Chapter 3. For simplicity of exposition, we omit the dose-volume constraints here, although they can be incorporated in a manner similar to Section 6.2. We thus consider the nominal formulation

$$\max_{N,u} \ \alpha_0 N(\bar{A}u) + \beta_0 N(\bar{A}u)^2 - \tau(N)$$

$$N\left[(A^m_{j.}u) + \rho_m (A^m_{j.}u)^2\right] \leq T_m \delta_m (1 + \rho_m \delta_m), \ j = 1 : n_m; \ m \in \mathcal{M}_1$$

$$N\left[\sum_{j=1}^{n_m} (A^m_{j.}u) + \rho_m (A^m_{j.}u)^2\right] \leq n_m T_m \delta_m (1 + \rho_m \delta_m), \ m \in \mathcal{M}_2$$

$$(6.14) - (6.17).$$

As in Chapter 4, we assume that OAR dose-response parameters ρ_m belong to the intervals $[\rho_m^{\min}, \rho_m^{\max}]$, for $m \in \mathcal{M}$. This yields the robust counterpart

$$(\text{RP}) \ \max_{N,u} \ \alpha_0 N(\bar{A}u) + \beta_0 N(\bar{A}u)^2 - \tau(N)$$

$$N\left[(A^m_{j.}u) + \rho_m (A^m_{j.}u)^2\right] \leq T_m \delta_m (1 + \rho_m \delta_m),$$

$$j = 1 : n_m; \ \forall \rho_m \in [\rho_m^{\min}, \rho_m^{\max}], \ m \in \mathcal{M}_1 \tag{7.1}$$

$$N\left[\sum_{j=1}^{n_m} (A^m_{j.}u) + \rho_m (A^m_{j.}u)^2\right] \leq n_m T_m \delta_m (1 + \rho_m \delta_m),$$

$$\forall \rho_m \in [\rho_m^{\min}, \rho_m^{\max}]; \ m \in \mathcal{M}_2 \tag{7.2}$$

$$(6.14) - (6.17)$$

of the above nominal formulation. Notice that (7.1) is a collection of an uncountably infinite number of constraints, and so is (7.2). In the next section, we describe an approach rooted in LP duality to reformulate this problem as a finite collection of finite-dimensional problems with linear objectives and convex quadratic constraints.

7.1 Tractable Reformulation

To construct a tractable reformulation of (RP), we first rewrite constraint (7.1) as

$$N(A^m_{j.}u) + \rho_m \left[N(A^m_{j.}u)^2 - T_m\delta_m^2\right] \leq T_m\delta_m, \tag{7.3}$$

by moving all terms that include ρ_m to the left-hand side. For any fixed (u, N) pair, this constraint holds if, and only if, the largest value of the left-hand side over all $\rho_m \in [\rho_m^{\min}, \rho_m^{\max}]$ is no more than the right-hand side. That is, if, and only if,

$$\left\{N(A^m_{j.}u) + \max_{\substack{\rho_m \geq \rho_m^{\min} \\ \rho_m \leq \rho_m^{\max}}} \rho_m \left[N(A^m_{j.}u)^2 - T_m\delta_m^2\right]\right\} \leq T_m\delta_m.$$

This is an LP with a single decision variable ρ_m bounded in the interval $[\rho_m^{\min}, \rho_m^{\max}]$. Thus, by strong duality, the optimal value of this LP is equal to the optimal value of its dual. To write this dual, we associate dual variable p_j^m with the primal constraint $\rho_m \geq \rho_m^{\min}$ and dual variable q_j^m with the primal constraint $\rho_m \leq \rho_m^{\max}$. Since $\rho_m^{\min} > 0$, the constraint $\rho_m \geq 0$ is implicit in the primal. We imagine it to be explicit while writing the dual. Thus, for any fixed (u, N) pair, constraint (7.3) holds if, and only if,

$$\left\{\begin{array}{c} N(A^m_{j.}u) + \min\limits_{p_j^m, q_j^m} \; p_j^m \rho_m^{\min} + q_j^m \rho_m^{\max} \\ p_j^m + q_j^m \geq N(A^m_{j.}u)^2 - T_m\delta_m^2 \\ p_j^m \leq 0, \; q_j^m \geq 0 \end{array}\right\} \leq T_m\delta_m. \tag{7.4}$$

Applying this logic again, one can show that constraint (7.2) holds if, and only if,

$$\left\{\begin{array}{c} N\sum_{j=1}^{n_m}(A^m_{j.}u) + \min\limits_{v^m, z^m} \; v^m \rho_m^{\min} + z^m \rho_m^{\max} \\ v^m + z^m \geq N\sum_{j=1}^{n_m}(A^m_{j.}u)^2 - n_m T_m\delta_m^2 \\ v^m \leq 0, \; z^m \geq 0 \end{array}\right\} \leq n_m T_m\delta_m. \tag{7.5}$$

This discussion shows that problem (RP) can be reformulated as

$$\text{(RP1)} \quad \max_{N,u,p,q,v,z} \quad \alpha_0 N(\bar{A}u) + \beta_0 N(\bar{A}u)^2 - \tau(N)$$

$$\left\{ \begin{array}{c} N(A^m_{j.}u) + \min\limits_{p^m_j, q^m_j} \; p^m_j \rho^{\min}_m + q^m_j \rho^{\max}_m \\ p^m_j + q^m_j \geq N(A^m_{j.}u)^2 - T_m \delta^2_m \\ p^m_j \leq 0, \; q^m_j \geq 0 \end{array} \right\} \leq T_m \delta_m,$$

$$j = 1 : n_m; \; m \in \mathcal{M}_1 \tag{7.6}$$

$$\left\{ \begin{array}{c} N \sum_{j=1}^{n_m} (A^m_{j.}u) + \min\limits_{v^m, z^m} \; v^m \rho^{\min}_m + z^m \rho^{\max}_m \\ v^m + z^m \geq N \sum_{j=1}^{n_m} (A^m_{j.}u)^2 - n_m T_m \delta^2_m \\ v^m \leq 0, \; z^m \geq 0 \end{array} \right\} \leq n_m T_m \delta_m, \; m \in \mathcal{M}_2 \tag{7.7}$$

$$(6.14) - (6.17).$$

This problem can be further simplified by dropping the minimization on the left-hand sides of (7.6) and (7.7) to obtain

$$\text{(RP2)} \quad \max_{N,u,p,q,v,z} \quad \alpha_0 N(\bar{A}u) + \beta_0 N(\bar{A}u)^2 - \tau(N)$$

$$N(A^m_{j.}u) + p^m_j \rho^{\min}_m + q^m_j \rho^{\max}_m \leq T_m \delta_m, \; j = 1 : n_m; \; m \in \mathcal{M}_1 \tag{7.8}$$

$$p^m_j + q^m_j \geq N(A^m_{j.}u)^2 - T_m \delta^2_m, \; j = 1 : n_m; \; m \in \mathcal{M}_1 \tag{7.9}$$

$$p^m_j \leq 0, \; q^m_j \geq 0, \; j = 1 : n_m; \; m \in \mathcal{M}_1 \tag{7.10}$$

$$N \sum_{j=1}^{n_m} (A^m_{j.}u) + v^m \rho^{\min}_m + z^m \rho^{\max}_m \leq n_m T_m \delta_m, \; m \in \mathcal{M}_2 \tag{7.11}$$

$$v^m + z^m \geq N \sum_{j=1}^{n_m} (A^m_{j.}u)^2 - n_m T_m \delta^2_m, \; m \in \mathcal{M}_2 \tag{7.12}$$

$$v^m \leq 0, \; z^m \geq 0, \; m \in \mathcal{M}_2 \tag{7.13}$$

$$(6.14) - (6.17). \tag{7.14}$$

The equivalence between the above two formulations can be seen as follows. If $N^\star, u^\star, p^\star, q^\star, v^\star, z^\star$ is an optimal solution to (RP1), then this solution is feasible to (RP1) with an identical objective value. Conversely, if $N^\star, u^\star, p^\star, q^\star, v^\star, z^\star$ is an optimal solution to (RP2), then a feasible solution of the form $N^\star, u^\star, \hat{p}, \hat{q}, \hat{v}, \hat{z}$ can be constructed for (RP2). This feasible solution has identical objective values in (RP1) and (RP2). Here, $\hat{p}, \hat{q}, \hat{v}, \hat{z}$ achieve the minima in the LPs that are embedded on the left-hand sides of constraints (7.6) and (7.7) in (RP1).

We first solve (RP2) for each fixed integer value of $1 \leq N \leq N_{\max}$ and then compare the resulting optimal objective values to find an optimal number of treatment sessions and a corresponding optimal intensity profile for (RP). As before, we note that when N is fixed, maximizing $\alpha_0 N(\bar{A}u) + \beta_0 N(\bar{A}u)^2 - \tau(N)$ is equivalent to maximizing $\bar{A}u$. Thus, the overall procedure involves solving a set of $N_{\max}$ problems in variables u, p, q, v, z. Each of these problems includes a linear objective and convex quadratic and linear constraints. These problems, indexed by N, are given by

$$\text{(RP2}(N)) \quad \max_{u,p,q,v,z} \quad \bar{A}u$$

$$(7.8) - (7.13)$$

$$(6.14) - (6.16).$$

This approach is summarized in Algorithm 5.

Algorithm 5 Exact solution of (RP)

1: Input - $n \times k$ tumor dose-deposition matrix A; tumor parameters $\alpha_0, \beta_0, T_{\text{double}}; N_{\max}$.
2: Input - set $\mathcal{M}$ of OAR, with mutually exclusive and exhaustive subsets $\mathcal{M}_1, \mathcal{M}_2$.
3: Input - $n_m \times k$ OAR dose-deposition matrices A^m, for each $m \in \mathcal{M}$.
4: Input - $T_m, \delta_m, [\rho_m^{\min}, \rho_m^{\max}]$, for each $m \in \mathcal{M}$.
5: Input - smoothness parameter $0 < \epsilon \leq 1$; beamlet adjacency matrix I.
6: Construct smoothness matrix S_ϵ.
7: best$= -\infty$.
8: **for** $N = 1 : N_{\max}$ **do**
9: Solve problem (RP2(N)) to obtain solution u^N.
10: Let $g^\star(N) = N\alpha_0(\bar{A}u^N) + N\beta_0(\bar{A}u^N)^2 - \tau(N)$ be the objective in (RP2).
11: **if** $g^\star(N) >$best **then** ▷ A better objective value found.
12: best$= g^\star(N)$.
13: $N_{\text{optimal}} = N$.
14: **end if**
15: **end for**
16: Output - $N_{\text{optimal}}, u^{N_{\text{optimal}}}$ as an optimal solution to (RP).

7.2 Numerical Experiments

We again consider the L-shaped example from Section 6.3. We fix tumor dose-response parameters at $\alpha_0 = 0.35$ Gy^{-1} and $\beta_0 = 0.035$ Gy^{-1}. We also fix

Table 7.1. *$N_{optimal}$ and percentage price of robustness for different Δ and T_{double}, when the single OAR is modeled as a serial one. The smoothness parameter was fixed at $\epsilon = 0.01$.*

		$\rho_1 = 1/5$ Gy^{-1}					
		T_{double} (days)					
		2	3	4	6	8	10
Δ	0	(5, 0%)	(7, 0%)	(10, 0%)	(14, 0%)	(18, 0%)	(22, 0%)
	0.1	(5, 2.88%)	(8, 2.24%)	(11, 1.81%)	(16, 1.24%)	(21, 0.87%)	(25, 0.59%)
	0.2	(6, 5.45%)	(9, 4.18%)	(12, 3.33%)	(17, 2.21%)	(22, 1.48%)	(27, 0.98%)
	0.3	(6, 7.56%)	(9, 5.90%)	(13, 4.63%)	(19, 3.05%)	(24, 2.01%)	(29, 1.25%)
	0.4	(6, 9.51%)	(10, 7.32%)	(13, 5.79%)	(20, 3.74%)	(26, 2.42%)	(31, 1.38%)
	0.5	(7, 11.22%)	(11, 8.59%)	(14, 6.73%)	(21, 4.29%)	(27, 2.69%)	(33, 1.51%)
	0.6	(7, 12.70%)	(11, 9.71%)	(15, 7.60%)	(22, 4.77%)	(29, 2.89%)	(35, 1.51%)
	0.7	(7, 14.03%)	(11, 10.75%)	(16, 8.39%)	(23, 5.19%)	(30, 3.09%)	(35, 1.51%)
	0.8	(7, 15.28%)	(12, 11.72%)	(16, 9.04%)	(24, 5.54%)	(32, 3.16%)	(35, 1.51%)
	0.9	(8, 16.44%)	(12, 12.54%)	(17, 9.70%)	(25, 5.88%)	(33, 3.22%)	(35, 1.51%)

the conventional number of treatment sessions and the tolerance dose for the single OAR at $T_1 = 35$ and $\delta_1 = 1.2857$ Gy. As in Section 6.3, we fix $\rho_1 = 1/5$ Gy^{-1}. However, recall, as in Section 4.3, that the algorithm does not know and does not utilize this nominal value. The nominal value is only employed to construct uncertainty intervals of the form $\rho_1^{\min} = (1-\Delta)\rho_1$ and $\rho_1^{\max} = (1+\Delta)\rho_1$, with $\Delta \in \{0, 0.1, 0.2, \ldots, 0.9\}$. Recall again that $\Delta = 0$ recovers the nominal problem, and the larger the Δ, the larger the uncertainty interval. Results for smoothness parameter $\epsilon = 0.01$ are displayed in Table 7.1, where the single OAR was modeled as a serial one. The qualitative trends in this table are consistent with Table 4.1, as expected.

Bibliographic Notes

This chapter is based on the doctoral dissertation [4] and an associated paper [6]. We refer the reader to that paper for extensive numerical experiments.

Exercises

Exercise 7.1 Repeat the numerical experiments in Table 7.1 for $\rho_1 = 1/4$ Gy^{-1} and $\rho_1 = 1/3$ Gy^{-1} and describe any qualitative trends you notice. Then, repeat all experiments for smoothness parameters $\epsilon = 0.1, 1$ and describe any qualitative trends you notice.

Exercise 7.2 Repeat all numerical experiments for the case where the single OAR is a parallel one with a mean-BED constraint. Describe any qualitative trends you notice.

Exercise 7.3 *Explore whether or not the robust optimization methodology in this chapter can be extended to the case where the anatomy includes a parallel OAR with dose-volume-BED constraints as in Chapter 6. In particular, attempt to design an approximate solution algorithm for the robust counterpart.

8

Fractionation with Two Modalities

In this chapter, we will generalize the model and solution method described in Chapter 3 to the case where two competing radiation modalities are available. For concreteness, we think of the first modality as the conventional one, that is, photon X-rays and the second modality as protons. We note, however, that our methodology does not in any way rely on these specific choices, and as such it can be applied to other pairs of modalities.

We wish to optimize the number of treatment sessions and the corresponding dosing plan administered with each modality. Unlike Chapter 3, we make the simplifying assumption that an equal-dosage solution is employed for each individual modality. Thus, the decision variables are N_1, d_1 for the first modality and N_2, d_2 for the second modality. All parameters for the first and the second modalities are denoted with subscripts 1 and 2, respectively. The total number of sessions is bounded above by $N_{\max}$, as before. Define $B_m = T_{1m}\delta_{1m}(\alpha_{1m} + \beta_{1m}\delta_{1m})$, for $m \in \mathcal{M}$. This yields the formulation

$$(2Q) \quad \max_{N_1, d_1, N_2, d_2} \sum_{i=1,2} N_i \left(\alpha_i^0 d_i + \beta_i^0 (d_i)^2\right) - \tau(N_1 + N_2) \tag{8.1}$$

$$\sum_{i=1,2} N_i \left(\alpha_{im}(s_{im} d_i) + \beta_{im}(s_{im} d_i)^2\right) \leq B_m, \; m \in \mathcal{M} \tag{8.2}$$

$$d_1 \geq 0, \; d_2 \geq 0 \tag{8.3}$$

$$0 \leq N_1, \text{ integer}; \; 0 \leq N_2, \text{ integer} \tag{8.4}$$

$$1 \leq N_1 + N_2 \leq N_{\max}. \tag{8.5}$$

Here, the objective function (8.1) equals the combined biological effect of the two modalities on the tumor, including the proliferation term $\tau(N_1 + N_2)$ that depends on the total number of treatment sessions $N_1 + N_2$. Note that we have included a 0 as a superscript to denote the dose-response parameters of the tumor, instead of including it as a subscript as in Chapter 3. This objective equals the negative of the natural logarithm of the surviving fraction of tumor cells. Hence, maximizing this objective is equivalent to minimizing this surviving fraction. The treatment planner's objective is thus identical to that in Chapter 3. The left-hand side of constraint (8.2) equals the combined biological effect of the two modalities on OAR $m \in \mathcal{M}$. In this left-hand side, α_{im}, β_{im} denote the dose-response parameters of OAR m for modalities $i = 1, 2$. Similarly, s_{im} are the sparing factors of OAR m for modalities $i = 1, 2$. The right-hand side of this constraint equals the biological effect on OAR m of a conventional treatment plan that includes T_{1m} sessions each with dose δ_{1m} administered via the first modality. The constraint thus ensures that the biological effect of the left-hand side is no more than this conventional value, which the OAR is known to tolerate. As such, this constraint is the counterpart of constraint (3.2) in Chapter 3. Observe, however, that both sides of the inequality in this constraint are unitless, unlike constraint (3.2) in Chapter 3, which was expressed in the units of radiation dose. Constraint (3.2) can indeed be rendered unitless by multiplying through with α_m, but that is not the standard practice in the literature. Conversely, constraint (8.2) can be converted into units of radiation dose, say by dividing through with α_{1m}. However, this is not advisable, since it would introduce ratios of dose-response parameters of different modalities into the formulation.

It is natural to solve problem (2Q) for each fixed integer pair of values (N_1, N_2) such that $N_1 \geq 0$, $N_2 \geq 0$, and $1 \leq N_1 + N_2 \leq N_{\max}$. We denote these problems by $(2Q(N_1, N_2))$. The resulting objective values of optimal solutions to these problems $(2Q(N_1, N_2))$ can then be compared to find an optimal pair of treatment sessions and a corresponding pair of optimal doses per session for problem (2Q). We present two analytical solution methods to solve problems $(2Q(N_1, N_2))$. Our discussion mostly focuses on the case where both N_1 and N_2 are positive integers. For if one of them is not, then the dose administered via that particular modality has no effect. That dose can therefore be arbitrarily set to 0 and the other dose can be calculated via solutions of quadratic equations as in the case of a single modality.

8.1 Analytical Solution Using KKT Conditions

For ease of exposition, we first focus on the case of a single OAR. We thus denote the corresponding optimization problems as (2Q1) and $(2Q1(N_1, N_2))$ and temporarily drop the subscript m from the dose-response parameters and sparing factors in our discussion. That is, we consider the problem

$$(2Q1(N_1,N_2)) \max_{d_1,d_2} \sum_{i=1,2} N_i\left(\alpha_i^0 d_i + \beta_i^0 (d_i)^2\right) - \tau(N_1+N_2) \tag{8.6}$$

$$\sum_{i=1,2} N_i \left(\alpha_i (s_i d_i) + \beta_i (s_i d_i)^2\right) \leq \underbrace{T_1\delta_1 (\alpha_1 + \beta_1\delta_1)}_{=B} \tag{8.7}$$

$$d_1 \geq 0, d_2 \geq 0. \tag{8.8}$$

Since the objective function is increasing in both d_1 and d_2, constraint (8.7) must be active at an optimal solution. For if not, we could increase either d_1 or d_2 and further increase the objective. There are three possibilities for optimal solutions: (i) $d_1 > 0, d_2 = 0$; (ii) $d_1 = 0, d_2 > 0$; and (iii) $d_1 > 0, d_2 > 0$. The first two are called single-dosage solutions and the third one is called double-dosage.

In the first case, the value of d_1 can be derived by solving the quadratic equation implied by the active form of constraint (8.7). This yields

$$d_1 = \frac{-\alpha_1 + \sqrt{(\alpha_1)^2 + 4\beta_1 B/N_1}}{2\beta_1 s_1}. \tag{8.9}$$

Similarly, in the second case, we have

$$d_2 = \frac{-\alpha_2 + \sqrt{(\alpha_2)^2 + 4\beta_2 B/N_2}}{2\beta_2 s_2}. \tag{8.10}$$

To analyze the third case, we rely on the KKT necessary conditions for optimality. Specifically, we associate Lagrange multiplier λ with the active form of constraint (8.7) and Lagrange multipliers μ_1, μ_2 with the nonnegativity constraints in (8.8), respectively. The KKT conditions are then given by $\mu_1 \geq 0$, $\mu_2 \geq 0$, $\mu_1 d_1 = 0$, $\mu_2 d_2 = 0$, $d_1 \geq 0$, $d_2 \geq 0$, and

$$N_1 s_1 \alpha_1 d_1 + N_1 \beta_1 (s_1 d_1)^2 + N_2 s_2 \alpha_2 d_2 + N_2 \beta_2 (s_2 d_2)^2 = B, \tag{8.11}$$

$$\begin{bmatrix} -\mu_1 \\ -\mu_2 \end{bmatrix} + \lambda \begin{bmatrix} N_1 s_1 \alpha_1 + 2N_1\beta_1 s_1^2 d_1 \\ N_2 s_2 \alpha_2 + 2N_2 \beta_2 s_2^2 d_2 \end{bmatrix} = \begin{bmatrix} N_1 \alpha_1^0 + 2N_1 \beta_1^0 d_1 \\ N_2 \alpha_2^0 + 2N_2\beta_2^0 d_2 \end{bmatrix}. \tag{8.12}$$

Since $d_1 > 0$ and $d_2 > 0$, we know that $\mu_1 = \mu_2 = 0$. Substituting this into the system (8.12) of equations yields

$$d_1 = \frac{\lambda s_1 \alpha_1 - \alpha_1^0}{2(\beta_1^0 - \lambda \beta_1 s_1^2)}, \tag{8.13}$$

$$d_2 = \frac{\lambda s_2 \alpha_2 - \alpha_2^0}{2(\beta_2^0 - \lambda \beta_2 s_2^2)}. \tag{8.14}$$

We will later separately handle the cases where the denominators of these two expressions are 0. So, for the moment, putting that possibility aside, we substitute (8.13)–(8.14) back into (8.11) to obtain

$$\sum_{i=1,2} \left\{ N_i s_i \alpha_i \left(\frac{\lambda s_i \alpha_i - \alpha_i^0}{2(\beta_i^0 - \lambda \beta_i s_i^2)} \right) + N_i \beta_i s_i^2 \left(\frac{\lambda s_i \alpha_i - \alpha_i^0}{2(\beta_i^0 - \lambda \beta_i s_i^2)} \right)^2 \right\} = B.$$

The least common multiplier (LCM) of the denominators in the above equation is

$$4(\beta_1^0 - \lambda \beta_1 s_1^2)^2 (\beta_2^0 - \lambda \beta_2 s_2^2)^2.$$

Multiplying through by this LCM yields

$$\sum_{i=1,2} \left\{ N_i s_i \alpha_i \left((\lambda s_i \alpha_i - \alpha_i^0) 2(\beta_i^0 - \lambda \beta_i s_i^2)(\beta_{-i}^0 - \lambda \beta_{-i} s_{-i}^2)^2 \right) + \right.$$

$$\left. N_i \beta_i s_i^2 \left((\lambda s_i \alpha_i - \alpha_i^0)^2 (\beta_{-i}^0 - \lambda \beta_{-i} s_{-i}^2)^2 \right) \right\} - 4B \prod_{i=1}^{2} (\beta_i^0 - \lambda \beta_i s_i^2)^2 = 0.$$

In this equation, we have used the notation $-i$ to denote the index that is not i. That is, $i = 1$ implies $-i = 2$; and $i = 2$ implies $-i = 1$. We simplify this equation as follows. Factoring out the expression $N_i s_i (\lambda s_i \alpha_i - \alpha_i^0)(\beta_{-i}^0 - \lambda \beta_{-i} s_{-i}^2)^2$ in the summand and using the notation $Z = 2\alpha_i(\beta_i^0 - \lambda \beta_i s_i^2) + \beta_i s_i (\lambda s_i \alpha_i - \alpha_i^0)$ yields

$$\sum_{i=1,2} \left\{ N_i s_i (\lambda s_i \alpha_i - \alpha_i^0)(\beta_{-i}^0 - \lambda \beta_{-i} s_{-i}^2)^2 Z \right\} - 4B \prod_{i=1}^{2} (\beta_i^0 - \lambda \beta_i s_i^2)^2 = 0.$$

The expression that is denoted by Z above simplifies as

$$\begin{aligned} Z &= 2\alpha_i \beta_i^0 - 2\lambda \alpha_i \beta_i s_i^2 + \lambda \alpha_i \beta_i s_i^2 - \beta_i s_i \alpha_i^0 = 2\alpha_i \beta_i^0 - \lambda \alpha_i \beta_i s_i^2 - \beta_i s_i \alpha_i^0 \\ &= 2\alpha_i \beta_i^0 - \beta_i s_i (\lambda \alpha_i s_i + \alpha_i^0). \end{aligned}$$

Let $Y = (\lambda s_i \alpha_i - \alpha_i^0)(\beta_{-i}^0 - \lambda \beta_{-i} s_{-i}^2)^2 \left(2\alpha_i \beta_i^0 - \beta_i s_i (\lambda \alpha_i s_i + \alpha_i^0)\right)$. We substitute this back into the earlier equation to obtain

$$\sum_{i=1,2} N_i s_i Y - 4B \underbrace{\prod_{i=1}^{2} (\beta_i^0 - \lambda \beta_i s_i^2)^2}_{=X} = 0. \tag{8.15}$$

The expression that is denoted by Y above is expanded as

$$\begin{aligned} Y &= 2\alpha_i \beta_i^0 (\lambda s_i \alpha_i - \alpha_i^0)(\beta_{-i}^0 - \lambda \beta_{-i} s_{-i}^2)^2 \\ &\quad - \beta_i s_i (\lambda^2 s_i^2 \alpha_i^2 - (\alpha_i^0)^2)(\beta_{-i}^0 - \lambda \beta_{-i} s_{-i}^2)^2 \\ &= 2\alpha_i \beta_i^0 (\lambda s_i \alpha_i - \alpha_i^0)((\beta_{-i}^0)^2 + \lambda^2 (\beta_{-i})^2 s_{-i}^4 - 2\beta_{-i}^0 \lambda \beta_{-i} s_{-i}^2) \\ &\quad - \beta_i s_i (\lambda^2 s_i^2 \alpha_i^2 - (\alpha_i^0)^2)((\beta_{-i}^0)^2 + \lambda^2 (\beta_{-i})^2 s_{-i}^4 - 2\beta_{-i}^0 \lambda \beta_{-i} s_{-i}^2) \end{aligned}$$

$$
\begin{aligned}
&= 2\alpha_i\beta_i^0(\lambda s_i\alpha_i(\beta_{-i}^0)^2 + \lambda^3 s_i\alpha_i(\beta_{-i})^2 s_{-i}^4 - 2\lambda^2 s_i\alpha_i\beta_{-i}^0\beta_{-i}s_{-i}^2)\\
&\quad - 2\alpha_i\beta_i^0\alpha_i^0((\beta_{-i}^0)^2 + \lambda^2(\beta_{-i})^2 s_{-i}^4 - 2\lambda\beta_{-i}^0\beta_{-i}s_{-i}^2)\\
&\quad - \beta_i s_i(\lambda^2 s_i^2\alpha_i^2(\beta_{-i}^0)^2 + \lambda^4 s_i^2\alpha_i^2(\beta_{-i})^2 s_{-i}^4 - 2\lambda^3 s_i^2\alpha_i^2\beta_{-i}^0\beta_{-i}s_{-i}^2)\\
&\quad + \beta_i s_i(\alpha_i^0)^2((\beta_{-i}^0)^2 + \lambda^2(\beta_{-i})^2 s_{-i}^4 - 2\beta_{-i}^0\lambda\beta_{-i}s_{-i}^2)\\
&= \lambda^4\left[-s_i^3\alpha_i^2\beta_i(\beta_{-i})^2 s_{-i}^4\right]\\
&\quad + \lambda^3\left[2s_i\beta_i^0\alpha_i^2(\beta_{-i})^2 s_{-i}^4 + 2s_i^3\alpha_i^2\beta_i\beta_{-i}^0\beta_{-i}s_{-i}^2\right]\\
&\quad + \lambda^2\big[-4s_i\beta_i^0\alpha_i^2\beta_{-i}^0\beta_{-i}s_{-i}^2 - 2\alpha_i^0\beta_i^0\alpha_i(\beta_{-i})^2 s_{-i}^4 - \alpha_i^2\beta_i s_i^3(\beta_{-i}^0)^2\\
&\qquad + s_i(\alpha_i^0)^2\beta_i(\beta_{-i})^2 s_{-i}^4\big]\\
&\quad + \lambda\left[2s_i\beta_i^0\alpha_i^2(\beta_{-i}^0)^2 + 4\alpha_i\beta_i^0\alpha_i^0\beta_{-i}^0\beta_{-i}s_{-i}^2 - 2\beta_i s_i(\alpha_i^0)^2\beta_{-i}^0\beta_{-i}s_{-i}^2\right]\\
&\qquad - 2\alpha_i^0\beta_i^0\alpha_i(\beta_{-i}^0)^2 + \beta_i s_i(\alpha_i^0)^2(\beta_{-i}^0)^2.
\end{aligned}
$$

The expression that is denoted X in (8.15) is expanded as

$$
\begin{aligned}
X &= ((\beta_1^0)^2 + \lambda^2\beta_1^2 s_1^4 - 2\lambda\beta_1^0\beta_1 s_1^2)((\beta_2^0)^2 + \lambda^2\beta_2^2 s_2^4 - 2\lambda\beta_2^0\beta_2 s_2^2)\\
&= \lambda^4\left[\beta_1^2 s_1^4\beta_2^2 s_2^4\right] + \lambda^3\left[-2\beta_1^2 s_1^4\beta_2^0\beta_2 s_2^2 - 2\beta_1^0\beta_1 s_1^2\beta_2^2 s_2^4\right]\\
&\quad + \lambda^2\left[(\beta_1^0)^2\beta_2^2 s_2^4 + \beta_1^2 s_1^4(\beta_2^0)^2 + 4\beta_1^0\beta_1 s_1^2\beta_2^0\beta_2 s_2^2\right]\\
&\quad + \lambda\left[-2(\beta_1^0)^2\beta_2^0\beta_2 s_2^2 - 2\beta_1^0\beta_1 s_1^2(\beta_2^0)^2\right] + (\beta_1^0)^2(\beta_2^0)^2.
\end{aligned}
$$

We further rewrite this as a symmetric sum by splitting some of the terms into two identical ones. For instance, we write the coefficient of λ^4, which is $\beta_1^2 s_1^4\beta_2^2 s_2^4$, as $0.5\beta_1^2 s_1^4\beta_2^2 s_2^4 + 0.5\beta_1^2 s_1^4\beta_2^2 s_2^4$. This yields

$$
\begin{aligned}
X = \sum_{i=1,2}\Big\{&\lambda^4\left[0.5\beta_i^2 s_i^4\beta_{-i}^2 s_{-i}^4\right] + \lambda^3\left[-2\beta_i^2 s_i^4\beta_{-i}^0\beta_{-i}s_{-i}^2\right]\\
&+ \lambda^2\left[(\beta_i^0)^2\beta_{-i}^2 s_{-i}^4 + 2\beta_i^0\beta_i s_i^2\beta_{-i}^0\beta_{-i}s_{-i}^2\right]\\
&+ \lambda\left[-2(\beta_i^0)^2\beta_{-i}^0\beta_{-i}s_{-i}^2\right] + 0.5(\beta_i^0)^2(\beta_{-i}^0)^2\Big\}.
\end{aligned}
$$

Now putting this expression for X and the above expression for Y back into (8.15) yields a quartic equation of the form

$$
c_0\lambda^4 + c_1\lambda^3 + c_2\lambda^2 + c_3\lambda + c_4 = 0 \tag{8.16}
$$

in λ. Here, the coefficients are given by

$$c_0 = \sum_{i=1,2} \Big\{ - N_i s_i^4 \alpha_i^2 \beta_i (\beta_{-i})^2 s_{-i}^4 - 2B\beta_i^2 s_i^4 \beta_{-i}^2 s_{-i}^4 \Big\}, \tag{8.17}$$

$$c_1 = \sum_{i=1,2} \Big\{ 2N_i s_i^2 \beta_i^0 \alpha_i^2 (\beta_{-i})^2 s_{-i}^4 + 2N_i s_i^4 \alpha_i^2 \beta_i \beta_{-i}^0 \beta_{-i} s_{-i}^2 + 4B\beta_i^2 s_i^4 \beta_{-i}^0 \beta_{-i} s_{-i}^2 \Big\}, \tag{8.18}$$

$$c_2 = \sum_{i=1,2} \Big\{ - 4N_i s_i^2 \beta_i^0 \alpha_i^2 \beta_{-i}^0 \beta_{-i} s_{-i}^2 - 2N_i s_i \alpha_i^0 \beta_i^0 \alpha_i (\beta_{-i})^2 s_{-i}^4 - N_i \alpha_i^2 \beta_i s_i^4 (\beta_{-i}^0)^2 + N_i s_i^2 (\alpha_i^0)^2 \beta_i (\beta_{-i})^2 s_{-i}^4 - 4B(\beta_i^0)^2 \beta_{-i}^2 s_{-i}^4 - 8B\beta_i^0 \beta_i s_i^2 \beta_{-i}^0 \beta_{-i} s_{-i}^2 \Big\}, \tag{8.19}$$

$$c_3 = \sum_{i=1,2} \Big\{ 2N_i s_i^2 \beta_i^0 \alpha_i^2 (\beta_{-i}^0)^2 + 4N_i s_i \alpha_i \beta_i^0 \alpha_i^0 \beta_{-i}^0 \beta_{-i} s_{-i}^2 - 2N_i \beta_i s_i^2 (\alpha_i^0)^2 \beta_{-i}^0 \beta_{-i} s_{-i}^2 + 8B(\beta_i^0)^2 \beta_{-i}^0 \beta_{-i} s_{-i}^2 \Big\}, \tag{8.20}$$

and

$$c_4 = \sum_{i=1,2} \Big\{ - 2N_i s_i \alpha_i^0 \beta_i^0 \alpha_i (\beta_{-i}^0)^2 + N_i \beta_i s_i^2 (\alpha_i^0)^2 (\beta_{-i}^0)^2 - 2B(\beta_i^0)^2 (\beta_{-i}^0)^2 \Big\}. \tag{8.21}$$

This quartic equation can be solved analytically using standard algebraic methods. By the fundamental theorem of algebra and the complex conjugate root theorem, there are three possibilities: two of the solutions are real and two are imaginary (complex conjugate); all four solutions are real; and all four solutions are imaginary (two pairs of complex conjugates). Each solution provides a potential opportunity to substitute it back into (8.13)–(8.14) and thereby obtain real positive values of doses d_1, d_2. However, this process could render the denominator and/or the numerator in (8.13) zero. Similarly, the denominator and/or the numerator in (8.14) could become zero. We thus carefully consider several subcases to identify the ones that are algebraically meaningful, in the next section.

8.1.1 Recovering Doses from a Real Solution of the Quartic

First suppose that $\lambda^\star$ is an imaginary solution of the above quartic equation. Then, the denominators of (8.13) and (8.14) cannot be 0; we can thus calculate

d_1 and d_2 from those two formulas. If those two values are real and positive, then they provide a candidate pair.

Now suppose that $\lambda^\star$ is a real solution of the above quartic equation. Recall that there can be either no such solutions, two such solutions, or four such solutions. We consider four cases depending on whether or not the denominator and/or the numerator of (8.13) is zero. Each case includes further subcases, depending on whether or not the denominator and/or the numerator of (8.14) is zero. We will utilize the shorthand notation

$$\Delta_i(d_{-i}) = 4N_i\beta_i s_i^2\left(B - N_{-i}s_{-i}\alpha_{-i}d_{-i} - N_{-i}\beta_{-i}s_{-i}^2 d_{-i}^2\right), \qquad (8.22)$$

for $i \in \{1,2\}$, in our analysis below.

Case 1 Suppose $\lambda^\star = \beta_1^0/(\beta_1 s_1^2) = \alpha_1^0/(s_1\alpha_1)$. That is, the denominator and numerator of (8.13) are zero.

1. Suppose $\lambda^\star = \beta_2^0/(\beta_2 s_2^2)$. That is, the denominator of (8.14) is zero.

   (i) Suppose $\lambda^\star = \alpha_2^0/(\alpha_2 s_2)$. That is, the numerator of (8.14) is zero.
   This means that $\frac{\alpha_1^0}{\alpha_1 s_1} = \frac{\alpha_2^0}{\alpha_2 s_2} = \frac{\beta_1^0}{\beta_1 s_1^2} = \frac{\beta_2^0}{\beta_2 s_2^2} = c$, for some constant c. Problem (2Q1(N_1, N_2)) can thus be rewritten by replacing α_i^0 with $\alpha_i s_i c$ and β_i^0 with $\beta_i s_i^2 c$, for $i = 1,2$. This yields

$$\max\ cB - \tau(N_1 + N_2)$$
$$N_1 s_1\alpha_1 d_1 + N_1\beta_1(s_1 d_1)^2 + N_2 s_2\alpha_2 d_2 + N_2\beta_2(s_2 d_2)^2 = B$$
$$d_1, d_2 \geq 0.$$

   Thus, all feasible solutions are optimal as the objective function equals a constant (a representative feasible solution can be obtained, for example, by setting $d_2 = 0$ and then obtaining d_1 by solving a quadratic equation to yield (8.9)).

   (ii) Suppose $\lambda^\star \neq \alpha_2^0/(\alpha_2 s_2)$. That is, the numerator of (8.14) is not zero.
   This does not yield a feasible d_2, because the denominator of (8.14) is zero but the numerator is not.

2. Suppose $\lambda^\star \neq \beta_2^0/(\beta_2 s_2^2)$. That is, the denominator of (8.14) is not zero.

(i) Suppose $\lambda^\star \neq \alpha_2^0/(\alpha_2 s_2)$. That is, the numerator of (8.14) is not zero.
Calculate d_2 from (8.14). If $N_2 s_2 \alpha_2 d_2 + N_2 \beta_2 s_2^2 d_2^2 \leq B$, solve the quadratic equation (8.11) to obtain d_1. That is,

$$d_1 = \frac{-N_1 s_1 \alpha_1 + \sqrt{N_1^2 s_1^2 (\alpha_1)^2 + \Delta_1(d_2)}}{2N_1 \beta_1 s_1^2}. \tag{8.23}$$

If $N_2 s_2 \alpha_2 d_2 + N_2 \beta_2 s_2^2 d_2^2 > B$, then this case does not yield a candidate pair.

(ii) Suppose $\lambda^\star = \alpha_2^0/(\alpha_2 s_2)$. That is, the numerator of (8.14) is zero.
This implies from (8.14) that $d_2 = 0$. This contradicts the assumption that $d_2 > 0$.

Case 2 Suppose $\lambda^\star = \beta_1^0/(\beta_1 s_1^2)$ and $\lambda^\star \neq \alpha_1^0/(\alpha_1 s_1)$. That is, the denominator of (8.13) is zero but the numerator is not.
This does not yield a feasible d_1 from (8.13).

Case 3 Suppose $\lambda^\star \neq \beta_1^0/(\beta_1 s_1^2)$ and $\lambda^\star = \alpha_1^0/(\alpha_1 s_1)$. That is, the denominator of (8.13) is not zero but the numerator is.
This implies from (8.13) that $d_1 = 0$. This contradicts the assumption that $d_1 > 0$ and $d_2 > 0$.

Case 4 Suppose $\lambda^\star \neq \beta_1^0/(\beta_1 s_1^2)$ and $\lambda^\star \neq \alpha_1^0/(\alpha_1 s_1)$. Neither the denominator nor the numerator of (8.13) is zero.

1. Suppose $\lambda^\star = \beta_2^0/(\beta_2 s_2^2)$. That is, the denominator of (8.14) is zero.

(i) Suppose $\lambda^\star = \alpha_2^0/(\alpha_2 s_2)$. That is, the numerator of (8.14) is zero.
Calculate d_1 from (8.13). If $N_1 s_1 \alpha_1 d_1 + N_1 \beta_1 s_1^2 d_1^2 \leq B$, solve the quadratic equation (8.11) to obtain d_2. That is,

$$d_2 = \frac{-N_2 s_2 \alpha_2 + \sqrt{N_2^2 s_2^2 (\alpha_2)^2 + \Delta_2(d_1)}}{2N_2 \beta_2 (s_2)^2}. \tag{8.24}$$

If $N_1 s_1 \alpha_1 d_1 + N_1 \beta_1 s_1^2 d_1^2 > B$, then this case does not yield a candidate pair.

(ii) Suppose $\lambda^\star \neq \alpha_2^0/(\alpha_2 s_2)$. That is, the numerator of (8.14) is not zero.
This does not yield a feasible d_2 from (8.14).

2. Suppose $\lambda^\star \neq \beta_2^0/(\beta_2 s_2^2)$. That is, the denominator of (8.14) is not zero.

   (i) Suppose $\lambda^\star \neq \alpha_2^0/(\alpha_2 s_2)$. That is, the numerator of (8.14) is not zero.
   In this case, d_1 and d_2 can be obtained from (8.13)–(8.14).

   (ii) Suppose $\lambda^\star = \alpha_2^0/(\alpha_2 s_2)$. That is, the numerator of (8.14) is zero.
   This implies from (8.14) that $d_2 = 0$. This contradicts the assumption that $d_2 > 0$ and thus does not yield a candidate pair.

In particular, these four cases yield one candidate pair of positive doses d_1, d_2 from each real solution $\lambda^\star$ (if any) of the quartic equation (8.16). Recall that these are called double-dosage candidates. We substitute each candidate d_1, d_2 pair (single-dosage from (8.9)–(8.10) and double-dosage from the four cases above) into (8.6) and choose a candidate pair with the largest objective value as an optimal solution to $(2\text{Q1}(N_1, N_2))$. The overall procedure is summarized in Algorithm 6.

8.2 Analytical Solution Based on a Single-Variable Reformulation

Recall that constraint (8.7) is active without loss of optimality in problem $(2\text{Q1}(N_1, N_2))$. Before describing our single-variable reformulation, we first tackle a special case where an optimal solution is evident. That case is thus excluded from our subsequent analysis.

Specifically, suppose that $\frac{\alpha_i^0}{\beta_i^0} - \frac{\alpha_i}{\beta_i s_i} = 0$, for $i = 1, 2$. We substitute $\alpha_1^0 = \beta_1^0 \frac{\alpha_1}{s_1 \beta_1}$ and $\alpha_2^0 = \beta_2^0 \frac{\alpha_2}{s_2 \beta_2}$ into the objective function of $(2\text{Q1}(N_1, N_2))$ to simplify it as

$$
\begin{aligned}
& N_1 \left(\beta_1^0 \frac{\alpha_1}{s_1 \beta_1} d_1 + \beta_1^0 d_1^2 \right) + N_2 \left(\beta_2^0 \frac{\alpha_2}{s_2 \beta_2} d_2 + \beta_2^0 d_2^2 \right) \\
&= \frac{N_1 \beta_1^0}{s_1 \beta_1} (\alpha_1 d_1 + \beta_1 s_1 d_1) + \frac{N_2 \beta_2^0}{s_2 \beta_2} (\alpha_2 d_2 + \beta_2 s_2 d_2) \\
&= \frac{N_1 \beta_1^0}{s_1^2 \beta_1} \left(\alpha_1 s_1 d_1 + \beta_1 s_1^2 d_1^2 \right) + \frac{N_2 \beta_2^0}{s_2^2 \beta_2} \left(\alpha_2 s_2 d_2 + \beta_2 s_2^2 d_2^2 \right).
\end{aligned}
$$

Algorithm 6 Exact solution of (2Q1) using KKT conditions

1: Input - α_i^0, β_i^0, for $i = 1,2$, T_{double}; α_i, β_i, s_i, for $i = 1,2$, and T_1, δ_1; $N_{\max}$.
2: Calculate $B = T_1\delta_1(\alpha_1 + \beta_1\delta_1)$.
3: best $= -\infty$.
4: **for** $N_1 = 0 : N_{\max}$ **do**
5: **for** $N_2 = \max\{0, 1 - N_1\} : N_{\max} - N_1$ **do** ▷ Ensures that $1 \le N_1 + N_2 \le N_{\max}$.
6: $\mathcal{L} = \{\}$. ▷ Empty list of candidate dose pairs (d_1, d_2).
7: **if** $N_1 = 0$ **then** ▷ It is guaranteed that $N_2 \neq 0$.
8: Let $d_1 = 0$ and $d_2 = \frac{-\alpha_2 + \sqrt{(\alpha_2)^2 + 4\beta_2 B/N_2}}{2\beta_2 s_2}$.
Append (d_1, d_2) to $\mathcal{L}$.
9: **else if** $N_2 = 0$ **then** ▷ It is guaranteed that $N_1 \neq 0$.
10: Let $d_1 = \frac{-\alpha_1 + \sqrt{(\alpha_1)^2 + 4\beta_1 B/N_1}}{2\beta_1 s_1}$ and $d_2 = 0$.
Append (d_1, d_2) to $\mathcal{L}$.
11: **else** ▷ $N_1 > 0$ and $N_2 > 0$.
12: Let $d_1 = \frac{-\alpha_1 + \sqrt{(\alpha_1)^2 + 4\beta_1 B/N_1}}{2\beta_1 s_1}$ and $d_2 = 0$.
Append (d_1, d_2) to $\mathcal{L}$. ▷ Single-dosage.
13: Let $d_1 = 0$ and $d_2 = \frac{-\alpha_2 + \sqrt{(\alpha_2)^2 + 4\beta_2 B/N_2}}{2\beta_2 s_2}$.
Append (d_1, d_2) to $\mathcal{L}$. ▷ Single-dosage.
14: Solve the quartic equation (8.16) for λ.
15: **for** Each real or imaginary solution $\lambda^\star$ of the quartic **do**
16: **if** $\lambda^\star$ is real **then**
17: Recover at most one pair (d_1, d_2) from $\lambda^\star$. ▷ Use Cases 1-4 from Section 8.1.1.
18: Append (d_1, d_2) to $\mathcal{L}$ if both are positive.
19: **else** ▷ $\lambda^\star$ is imaginary.
20: Calculate d_1, d_2 from (8.13) and (8.14).
21: Append (d_1, d_2) to $\mathcal{L}$ if both are real and positive.
22: **end if**
23: **end for**
24: **end if**
25: **for** Each pair (d_1, d_2) of doses in $\mathcal{L}$ **do**
26: Calculate obj $= \left[\sum_{i=1,2} N_i \left(\alpha_i^0 d_i + \beta_i^0 (d_i)^2\right)\right] - \tau(N_1 + N_2)$.
27: **if** obj > best **then** ▷ A better objective value found.
28: best = obj.
29: $N_1^{\text{optimal}} = N_1$; $d_1^{\text{optimal}} = d_1$;
$N_2^{\text{optimal}} = N_2$; and $d_2^{\text{optimal}} = d_2$.
30: **end if**
31: **end for**
32: **end for**
33: **end for**
34: Output - $N_1^{\text{optimal}}, d_1^{\text{optimal}}, N_2^{\text{optimal}} = N_2, d_2^{\text{optimal}} = d_2$.

Now substituting $\alpha_2 s_2 d_2 + \beta_2 s_2^2 d_2^2 = B - \alpha_1 s_1 d_1 + \beta_1 s_1^2 d_1^2$ into this expression yields

$$\frac{N_1 \beta_1^0}{s_1^2 \beta_1}\left(\alpha_1 s_1 d_1 + \beta_1 s_1^2 d_1^2\right) + \frac{N_2 \beta_2^0}{s_2^2 \beta_2}\left(B - \alpha_1 s_1 d_1 + \beta_1 s_1^2 d_1^2\right)$$
$$= \frac{N_2 \beta_2^0 B}{s_2^2 \beta_2} + \left(\alpha_1 s_1 d_1 + \beta_1 s_1^2 d_1^2\right)\left(\frac{N_1 \beta_1^0}{s_1^2 \beta_1} - \frac{N_2 \beta_2^0}{s_2^2 \beta_2}\right).$$

This means that problem (2Q1(N_1, N_2)) reduces to maximizing this objective over nonnegative doses d_1 such that $\alpha_1 s_1 d_1 + \beta_1 s_1^2 d_1^2 \leq B$. Thus, an optimal solution is readily derived as follows:

$$d_1 = 0,\ d_2 = \frac{-\alpha_2 + \sqrt{(\alpha_2)^2 + 4\beta_2 B/N_2}}{2\beta_2 s_2},\ \text{if} \left(\frac{N_1 \beta_1^0}{s_1^2 \beta_1} - \frac{N_2 \beta_2^0}{s_2^2 \beta_2}\right) < 0, \text{ and}$$

$$d_1 = \frac{-\alpha_1 + \sqrt{(\alpha_1)^2 + 4\beta_1 B/N_1}}{2\beta_1 s_1},\ d_2 = 0, \text{otherwise.}$$

Our analysis below thus excludes this easy case.

Consider an alternative decision variable, $0 \leq x \leq 1$, which equals the fraction of B (the right-hand side in constraint (8.7)) that is administered using the first modality. Consequently, a fraction $0 \leq 1 - x \leq 1$ of B is administered using the second modality. Specifically, we have $N_1\left(\alpha_1(s_1 d_1) + \beta_1(s_1 d_1)^2\right) = xB$ and $N_2\left(\alpha_2(s_2 d_2) + \beta_1(s_2 d_2)^2\right) = (1 - x)B$. By solving these two quadratic equations for nonnegative d_1 and d_2 in terms of x, we reformulate (2Q1(N_1, N_2)) as a single-variable problem. In particular, for $0 \leq x \leq 1$, we have

$$d_1(x) = \frac{-\alpha_1 + \sqrt{(\alpha_1)^2 + 4\beta_1(xB)/N_1}}{2\beta_1 s_1}, \text{ and} \tag{8.25}$$

$$d_2(x) = \frac{-\alpha_2 + \sqrt{(\alpha_2)^2 + 4\beta_2((1-x)B)/N_2}}{2\beta_2 s_2}. \tag{8.26}$$

Substituting $d_1(x), d_2(x)$ back into the objective of (2Q1(N_1, N_2)) yields the equivalent reformulation

$$(2\text{Q1}(N_1, N_2)) \quad \max_{0 \leq x \leq 1} f(x) = f_1(x) + f_2(x) =$$

$$N_1 \left\{ \alpha_1^0 \left(\frac{-\alpha_1 + \sqrt{(\alpha_1)^2 + 4\beta_1(xB)/N_1}}{2\beta_1 s_1} \right) \right.$$
$$\left. + \beta_1^0 \left(\frac{-\alpha_1 + \sqrt{(\alpha_1)^2 + 4\beta_1(xB)/N_1}}{2\beta_1 s_1} \right)^2 \right\} +$$

$$N_2\left\{\alpha_2^0\left(\frac{-\alpha_2+\sqrt{(\alpha_2)^2+4\beta_2((1-x)B)/N_2}}{2\beta_2 s_2}\right)\right.$$
$$\left.+\beta_2^0\left(\frac{-\alpha_2+\sqrt{(\alpha_2)^2+4\beta_2((1-x)B)/N_2}}{2\beta_2 s_2}\right)^2\right\}. \tag{8.27}$$

Here, we have used $f_1(x)$ and $f_2(x)$ to denote the tumor biological effect of the two modalities, respectively. That is, $f_1(x)$ denotes the first expression above, and $f_2(x)$ denotes the second. Thus, an optimal solution occurs either at $x = 0$, or at $x = 1$, or at one of the values of x where

$$\frac{df(x)}{dx} = \frac{df_1(x)}{dx} + \frac{df_2(x)}{dx} = 0.$$

To compute its derivative, we first expand $f_1(x)$ as

$$f_1(x) = \frac{N_1\beta_1^0}{2\beta_1 s_1}\left[\frac{\alpha_1^0}{\beta_1^0}\left(-\alpha_1+\sqrt{(\alpha_1)^2+4\beta_1(xB)/N_1}\right)\right.$$
$$\left.+\left(\frac{(\alpha_1)^2-2\alpha_1\sqrt{(\alpha_1)^2+4\beta_1(xB)/N_1}+(\alpha_1)^2+4\beta_1(xB)/N_1}{2\beta_1 s_1}\right)\right].$$

For brevity, we temporarily define $y_1(x) = (\alpha_1)^2 + 4\beta_1(xB)/N_1$ and collect some of the constant terms to rewrite as

$$\begin{aligned} f_1(x) &= \frac{N_1\beta_1^0}{2\beta_1 s_1}\left[\frac{\alpha_1^0}{\beta_1^0}\left(-\alpha_1+\sqrt{y_1(x)}\right)+\left(\frac{(\alpha_1)^2-2\alpha_1\sqrt{y_1(x)}+y_1(x)}{2\beta_1 s_1}\right)\right] \\ &= \frac{N_1\beta_1^0}{2\beta_1 s_1}\left[\alpha_1\left(\frac{\alpha_1}{2\beta_1 s_1}-\frac{\alpha_1^0}{\beta_1^0}\right)+\left(\frac{\alpha_1^0}{\beta_1^0}-\frac{\alpha_1}{\beta_1 s_1}\right)\sqrt{y_1(x)}+\frac{y_1(x)}{2\beta_1 s_1}\right]. \end{aligned}$$

This implies that

$$\begin{aligned} \frac{df_1(x)}{dx} &= \frac{N_1\beta_1^0}{2\beta_1 s_1}\left[\left(\frac{\alpha_1^0}{\beta_1^0}-\frac{\alpha_1}{\beta_1 s_1}\right)\frac{1}{2\sqrt{y_1(x)}}+\frac{1}{2\beta_1 s_1}\right]\frac{dy_1(x)}{d(x)} \\ &= \frac{N_1\beta_1^0}{2\beta_1 s_1}\left[\left(\frac{\alpha_1^0}{\beta_1^0}-\frac{\alpha_1}{\beta_1 s_1}\right)\frac{1}{2\sqrt{y_1(x)}}+\frac{1}{2\beta_1 s_1}\right]\frac{4\beta_1 B}{N_1} \\ &= \frac{\beta_1^0}{s_1}\left[\left(\frac{\alpha_1^0}{\beta_1^0}-\frac{\alpha_1}{\beta_1 s_1}\right)\frac{1}{\sqrt{y_1(x)}}+\frac{1}{\beta_1 s_1}\right]. \end{aligned}$$

Following similar algebra and letting $y_2(x) = (\alpha_2)^2 + 4\beta_2((1-x)B)/N_2$, we obtain

$$
\begin{aligned}
\frac{df_2(x)}{dx} &= \frac{N_2\beta_2^0}{2\beta_2 s_2}\left[\left(\frac{\alpha_2^0}{\beta_2^0} - \frac{\alpha_2}{\beta_2 s_2}\right)\frac{1}{2\sqrt{y_2(x)}} + \frac{1}{2\beta_2 s_2}\right]\left(-\frac{4\beta_2 B}{N_2}\right) \\
&= -\frac{\beta_2^0}{s_2}\left[\left(\frac{\alpha_2^0}{\beta_2^0} - \frac{\alpha_2}{\beta_2 s_2}\right)\frac{1}{\sqrt{y_2(x)}} + \frac{1}{\beta_2 s_2}\right].
\end{aligned}
$$

Thus, we are interested in values of x that satisfy

$$
\begin{aligned}
&\frac{\beta_1^0}{s_1}\left[\left(\frac{\alpha_1^0}{\beta_1^0} - \frac{\alpha_1}{\beta_1 s_1}\right)\frac{1}{\sqrt{y_1(x)}} + \frac{1}{\beta_1 s_1}\right] \\
&\quad = \frac{\beta_2^0}{s_2}\left[\left(\frac{\alpha_2^0}{\beta_2^0} - \frac{\alpha_2}{\beta_2 s_2}\right)\frac{1}{\sqrt{y_2(x)}} + \frac{1}{\beta_2 s_2}\right].
\end{aligned}
\tag{8.28}
$$

We consider three cases to solve this equation: (i) $\frac{\alpha_1^0}{\beta_1^0} - \frac{\alpha_1}{\beta_1 s_1} = 0$ and $\frac{\alpha_2^0}{\beta_2^0} - \frac{\alpha_2}{\beta_2 s_2} \neq 0$; (ii) $\frac{\alpha_1^0}{\beta_1^0} - \frac{\alpha_1}{\beta_1 s_1} \neq 0$ and $\frac{\alpha_2^0}{\beta_2^0} - \frac{\alpha_2}{\beta_2 s_2} = 0$; and (iii) $\frac{\alpha_1^0}{\beta_1^0} - \frac{\alpha_1}{\beta_1 s_1} \neq 0$ and $\frac{\alpha_2^0}{\beta_2^0} - \frac{\alpha_2}{\beta_2 s_2} \neq 0$.

In the first case, (8.28) reduces to

$$
\frac{\beta_2^0}{s_2}\left[\left(\frac{\alpha_2^0}{\beta_2^0} - \frac{\alpha_2}{\beta_2 s_2}\right)\frac{1}{\sqrt{y_2(x)}} + \frac{1}{\beta_2 s_2}\right] = \frac{\beta_1^0}{s_1}\frac{1}{\beta_1 s_1},
$$

which further yields

$$
\frac{1}{\sqrt{y_2(x)}} = \frac{\frac{\beta_1^0}{\beta_2^0}\frac{s_2}{s_1}\frac{1}{\beta_1 s_1} - \frac{1}{\beta_2 s_2}}{\frac{\alpha_2^0}{\beta_2^0} - \frac{\alpha_2}{\beta_2 s_2}}.
\tag{8.29}
$$

Thus, as long as $\frac{\beta_1^0}{\beta_2^0}\frac{s_2}{s_1}\frac{1}{\beta_1 s_1} \neq \frac{1}{\beta_2 s_2}$, we get

$$
y_2(x) = \frac{\left(\frac{\alpha_2^0}{\beta_2^0} - \frac{\alpha_2}{\beta_2 s_2}\right)^2}{\left(\frac{\beta_1^0}{\beta_2^0}\frac{s_2}{s_1}\frac{1}{\beta_1 s_1} - \frac{1}{\beta_2 s_2}\right)^2}.
$$

This means that

$$
(\alpha_2)^2 + \frac{4\beta_2((1-x)B)}{N_2} = \frac{\left(\frac{\alpha_2^0}{\beta_2^0} - \frac{\alpha_2}{\beta_2 s_2}\right)^2}{\left(\frac{\beta_1^0}{\beta_2^0}\frac{s_2}{s_1}\frac{1}{\beta_1 s_1} - \frac{1}{\beta_2 s_2}\right)^2},
$$

which in turn yields

$$x = 1 + \frac{N_2}{4\beta_2 B}\left[\alpha_2^2 - \frac{\left(\frac{\alpha_2^0}{\beta_2^0} - \frac{\alpha_2}{\beta_2 s_2}\right)^2}{\left(\frac{\beta_1^0}{\beta_2^0}\frac{s_2}{s_1}\frac{1}{\beta_1 s_1} - \frac{1}{\beta_2 s_2}\right)^2}\right]. \tag{8.30}$$

If this value of x is between 0 and 1, it provides a candidate solution. If $\frac{\beta_1^0}{\beta_2^0}\frac{s_2}{s_1}\frac{1}{\beta_1 s_1} = \frac{1}{\beta_2 s_2}$, this first case does not yield a candidate solution from formula (8.29).

The second case is tackled similarly. In particular, (8.28) reduces to

$$\frac{\beta_1^0}{s_1}\left[\left(\frac{\alpha_1^0}{\beta_1^0} - \frac{\alpha_1}{\beta_1 s_1}\right)\frac{1}{\sqrt{y_1(x)}} + \frac{1}{\beta_1 s_1}\right] = \frac{\beta_2^0}{s_2}\frac{1}{\beta_2 s_2},$$

which further yields

$$\frac{1}{\sqrt{y_1(x)}} = \frac{\frac{\beta_2^0}{\beta_1^0}\frac{s_1}{s_2}\frac{1}{\beta_2 s_2} - \frac{1}{\beta_1 s_1}}{\frac{\alpha_1^0}{\beta_1^0} - \frac{\alpha_1}{\beta_1 s_1}}.$$

Thus, as long as $\frac{\beta_2^0}{\beta_1^0}\frac{s_1}{s_2}\frac{1}{\beta_2 s_2} \neq \frac{1}{\beta_1 s_1}$, we get

$$y_1(x) = \frac{\left(\frac{\alpha_1^0}{\beta_1^0} - \frac{\alpha_1}{\beta_1 s_1}\right)^2}{\left(\frac{\beta_2^0}{\beta_1^0}\frac{s_1}{s_2}\frac{1}{\beta_2 s_2} - \frac{1}{\beta_1 s_1}\right)^2}.$$

This means that

$$(\alpha_1)^2 + \frac{4\beta_1 x B}{N_1} = \frac{\left(\frac{\alpha_1^0}{\beta_1^0} - \frac{\alpha_1}{\beta_1 s_1}\right)^2}{\left(\frac{\beta_2^0}{\beta_1^0}\frac{s_1}{s_2}\frac{1}{\beta_2 s_2} - \frac{1}{\beta_1 s_1}\right)^2},$$

which in turn yields

$$x = \frac{N_1}{4\beta_1 B}\left[\frac{\left(\frac{\alpha_1^0}{\beta_1^0} - \frac{\alpha_1}{\beta_1 s_1}\right)^2}{\left(\frac{\beta_2^0}{\beta_1^0}\frac{s_1}{s_s}\frac{1}{\beta_2 s_2} - \frac{1}{\beta_1 s_1}\right)^2} - \alpha_1^2\right]. \tag{8.31}$$

If this value of x is between 0 and 1, it provides a candidate solution. If $\frac{\beta_2^0}{\beta_1^0}\frac{s_1}{s_2}\frac{1}{\beta_2 s_2} = \frac{1}{\beta_1 s_1}$, this second case does not yield a candidate solution.

In the third case, we denote both expressions on the two sides of the equality sign in (8.28) by a real number λ. Furthermore, we temporarily introduce constants

$$a_i = \frac{\beta_i^0}{s_i}\left(\frac{\alpha_i^0}{\beta_i^0} - \frac{\alpha_i}{\beta_i s_i}\right) \text{ and } b_i = \frac{\beta_i^0}{s_i}\frac{1}{\beta_i s_i},\ i = 1,2. \tag{8.32}$$

Using this, we obtain $\lambda - b_i = \frac{a_i}{\sqrt{y_i(x)}}$, for $i = 1,2$. If $\lambda - b_i = 0$, there is no corresponding x here because $a_i \neq 0$ in the third case that is currently under consideration. If $\lambda - b_i \neq 0$, we use the definitions of $y_1(x)$ and $y_2(x)$ and some algebraic simplification to obtain

$$\frac{N_1}{\beta_1}\left[\frac{a_1^2}{(\lambda - b_1)^2} - \alpha_1^2\right] = 4xB \text{ and } \frac{N_2}{\beta_2}\left[\frac{a_2^2}{(\lambda - b_2)^2} - \alpha_2^2\right] = 4B - 4xB. \tag{8.33}$$

Adding these two equations gives

$$\frac{N_1}{\beta_1}\frac{a_1^2}{(\lambda - b_1)^2} + \frac{N_2}{\beta_2}\frac{a_2^2}{(\lambda - b_2)^2} = 4B + \frac{N_1}{\beta_1}\alpha_1^2 + \frac{N_2}{\beta_2}\alpha_2^2.$$

We define three further constants as

$$c_0 = 4B + \frac{N_1}{\beta_1}\alpha_1^2 + \frac{N_2}{\beta_2}\alpha_2^2; \text{ and } c_i = \frac{N_i a_i^2}{\beta_i},\ i = 1,2, \tag{8.34}$$

which yields $\frac{c_1}{(\lambda-b_1)^2} + \frac{c_2}{(\lambda-b_2)^2} = c_0$. After algebraic simplification, this results in a quartic equation of the form

$$w_0\lambda^4 + w_1\lambda^3 + w_2\lambda^2 + w_3\lambda + w_4 = 0, \tag{8.35}$$

where

$$w_0 = c_0 \tag{8.36}$$

$$w_1 = -2c_0(b_1 + b_2) \tag{8.37}$$

$$w_2 = c_0(b_1^2 + b_2^2 + 4b_1b_2) - (c_1 + c_2) \tag{8.38}$$

$$w_3 = 2\,(c_1b_2 + c_2b_1 - c_0b_1b_2(b_1 + b_2)) \tag{8.39}$$

$$w_4 = c_0b_1^2b_2^2 - c_1b_2^2 - c_2b_1^2. \tag{8.40}$$

We substitute each (real as well as complex) solution of this quartic back into the first equation from the pair in (8.33) to obtain x. If this x is real valued and satisfies $0 \leq x \leq 1$, then it provides a candidate solution.

Regardless of which of the above three cases the problem parameters satisfy, an optimal dosing plan can be obtained by comparing the objective values of all candidate solutions x (including those with $x = 0$ and $x = 1$). Suppose $0 \leq x^\star \leq 1$ is a candidate solution that attains the largest objective value. Then, optimal doses for the two modalities can be calculated by substituting $x = x^\star$ in (8.25)–(8.26). The overall procedure is summarized in Algorithm 7.

We conclude this section by providing further insights into a special case of parameter values. We begin with calculating derivatives of the function $f(x)$ at $x = 0$ and $x = 1$. We first note, by definitions, that

$$
\begin{aligned}
&y_1(0) = (\alpha_1)^2,\ y_1(1) = (\alpha_1)^2 + 4\beta_1 B/N_1,\\
&y_2(0) = (\alpha_2)^2 + 4\beta_2 B/N_2, \text{ and } y_2(1) = (\alpha_2)^2.
\end{aligned}
$$

These observations imply that

$$
\begin{aligned}
\frac{df(0)}{dx} &= \frac{\beta_1^0}{s_1}\left[\left(\frac{\alpha_1^0}{\beta_1^0} - \frac{\alpha_1}{\beta_1 s_1}\right)\frac{1}{\sqrt{y_1(0)}} + \frac{1}{\beta_1 s_1}\right]\\
&\quad - \frac{\beta_2^0}{s_2}\left[\left(\frac{\alpha_2^0}{\beta_2^0} - \frac{\alpha_2}{\beta_2 s_2}\right)\frac{1}{\sqrt{y_2(0)}} + \frac{1}{\beta_2 s_2}\right]\\
&= \frac{\beta_1^0}{s_1}\left[\left(\frac{\alpha_1^0}{\beta_1^0} - \frac{\alpha_1}{\beta_1 s_1}\right)\frac{1}{\alpha_1} + \frac{1}{\beta_1 s_1}\right]\\
&\quad - \frac{\beta_2^0}{s_2}\left[\left(\frac{\alpha_2^0}{\beta_2^0} - \frac{\alpha_2}{\beta_2 s_2}\right)\frac{1}{\sqrt{(\alpha_2)^2 + 4\beta_2 B/N_2}} + \frac{1}{\beta_2 s_2}\right],
\end{aligned}
$$

and

$$
\begin{aligned}
\frac{df(1)}{dx} &= \frac{\beta_1^0}{s_1}\left[\left(\frac{\alpha_1^0}{\beta_1^0} - \frac{\alpha_1}{\beta_1 s_1}\right)\frac{1}{\sqrt{y_1(1)}} + \frac{1}{\beta_1 s_1}\right]\\
&\quad - \frac{\beta_2^0}{s_2}\left[\left(\frac{\alpha_2^0}{\beta_2^0} - \frac{\alpha_2}{\beta_2 s_2}\right)\frac{1}{\sqrt{y_2(1)}} + \frac{1}{\beta_2 s_2}\right]\\
&= \frac{\beta_1^0}{s_1}\left[\left(\frac{\alpha_1^0}{\beta_1^0} - \frac{\alpha_1}{\beta_1 s_1}\right)\frac{1}{\sqrt{(\alpha_1)^2 + 4\beta_1 B/N_1}} + \frac{1}{\beta_1 s_1}\right]\\
&\quad - \frac{\beta_2^0}{s_2}\left[\left(\frac{\alpha_2^0}{\beta_2^0} - \frac{\alpha_2}{\beta_2 s_2}\right)\frac{1}{\alpha_2} + \frac{1}{\beta_2 s_2}\right].
\end{aligned}
$$

These formulas are utilized in the next lemma.

Lemma 8.1 *Suppose* $\frac{\alpha_1^0}{\beta_1^0} - \frac{\alpha_1}{\beta_1 s_1} \geq 0$ *and* $\frac{\alpha_2^0}{\beta_2^0} - \frac{\alpha_2}{\beta_2 s_2} \geq 0$, *but at least one of these two is strictly positive. Then, the objective function* $f(x)$ *of the maximization*

Algorithm 7 Exact solution of (2Q1) using a single-variable transformation

1: Steps 1–13 from Algorithm 6.
2: **if** $\frac{\alpha_1^0}{\beta_1^0} - \frac{\alpha_1}{\beta_1 s_1} = 0$ and $\frac{\alpha_2^0}{\beta_2^0} - \frac{\alpha_2}{\beta_2 s_2} = 0$ **then** ▷ Easy case.
3: **if** $\left(\frac{N_1\beta_1^0}{s_1^2\beta_1} - \frac{N_2\beta_2^0}{s_2^2\beta_2}\right) < 0$ **then**
4: Let $d_1 = 0$ and $d_2 = \frac{-\alpha_2+\sqrt{(\alpha_2)^2+4\beta_2 B/N_2}}{2\beta_2 s_2}$. Append (d_1, d_2) to $\mathcal{L}$.
5: **else** ▷ $\left(\frac{N_1\beta_1^0}{s_1^2\beta_1} - \frac{N_2\beta_2^0}{s_2^2\beta_2}\right) \geq 0$.
6: Let $d_1 = \frac{-\alpha_1+\sqrt{(\alpha_1)^2+4\beta_1 B/N_1}}{2\beta_1 s_1}$ and $d_2 = 0$. Append (d_1, d_2) to $\mathcal{L}$.
7: **end if**
8: **else if** $\frac{\alpha_1^0}{\beta_1^0} - \frac{\alpha_1}{\beta_1 s_1} = 0$ and $\frac{\alpha_2^0}{\beta_2^0} - \frac{\alpha_2}{\beta_2 s_2} \neq 0$ **then** ▷ Case (i).
9: **if** $\frac{\beta_1^0}{\beta_2^0}\frac{s_2}{s_1}\frac{1}{\beta_1 s_1} \neq \frac{1}{\beta_2 s_2}$ **then**
10: Let $x = 1 + \frac{N_2}{4\beta_2 B}\left[\alpha_2^2 - \frac{\left(\frac{\alpha_2^0}{\beta_2^0} - \frac{\alpha_2}{\beta_2 s_2}\right)^2}{\left(\frac{\beta_1^0}{\beta_2^0}\frac{s_2}{s_1}\frac{1}{\beta_1 s_1} - \frac{1}{\beta_2 s_2}\right)^2}\right]$.
11: Append $(d_1(x), d_2(x))$ to $\mathcal{L}$ if $0 \leq x \leq 1$.
12: **end if**
13: **else if** $\frac{\alpha_1^0}{\beta_1^0} - \frac{\alpha_1}{\beta_1 s_1} \neq 0$ and $\frac{\alpha_2^0}{\beta_2^0} - \frac{\alpha_2}{\beta_2 s_2} = 0$ **then** ▷ Case (ii).
14: **if** $\frac{\beta_2^0}{\beta_1^0}\frac{s_1}{s_2}\frac{1}{\beta_2 s_2} \neq \frac{1}{\beta_1 s_1}$ **then**
15: Let $x = \frac{N_1}{4\beta_1 B}\left[\frac{\left(\frac{\alpha_1^0}{\beta_1^0} - \frac{\alpha_1}{\beta_1 s_1}\right)^2}{\left(\frac{\beta_2^0}{\beta_1^0}\frac{s_1}{s_s}\frac{1}{\beta_2 s_2} - \frac{1}{\beta_1 s_1}\right)^2} - \alpha_1^2\right]$.
16: Append $(d_1(x), d_2(x))$ to $\mathcal{L}$ if $0 \leq x \leq 1$.
17: **end if**
18: **else** ▷ Case (iii): $\frac{\alpha_1^0}{\beta_1^0} - \frac{\alpha_1}{\beta_1 s_1} \neq 0$ and $\frac{\alpha_2^0}{\beta_2^0} - \frac{\alpha_2}{\beta_2 s_2} \neq 0$.
19: Solve the quartic equation (8.35) for λ.
20: **for** Each solution $\lambda^\star$ of the quartic **do**
21: **if** $\lambda^\star - b_1 \neq 0$ **then**
22: Let $x = \frac{N_1}{4\beta_1 B}\left[\frac{a_1^2}{(\lambda^\star - b_1)^2} - \alpha_1^2\right]$ from (8.33).
23: Append $(d_1(x), d_2(x))$ to $\mathcal{L}$ if x is real and $0 \leq x \leq 1$.
24: **end if**
25: **end for**
26: **end if**
27: Steps 19–29 from Algorithm 6.

problem $(2Q1(N_1, N_2))$ *is strictly concave over* $0 < x < 1$. *Thus, only the three possibilities listed below could occur.*

1. *The solution* $x = 0$ *is optimal because*

$$\frac{df(0)}{dx} \leq 0, \text{ and } \frac{df(1)}{dx} < 0.$$

   See Figure 8.1(a).
2. *The solution* $x = 1$ *is optimal because*

$$\frac{df(0)}{dx} > 0, \text{ and } \frac{df(1)}{dx} \geq 0.$$

   See Figure 8.1(b).
3. *Equation* (8.28) *has a unique solution* $x^\star$*; this solution satisfies* $0 < x^\star < 1$ *and is optimal. See Figure 8.1(c).*

Proof The fact that the three cases are mutually exhaustive is a direct consequence of strict concavity of the objective. Similarly for the actual conclusions in the three cases. The proof thus entirely focuses on establishing strict concavity. Building upon the derivative calculations from earlier in this section, we obtain

$$\begin{aligned}\frac{d^2 f(x)}{dx^2} &= \frac{d^2 f_1(x)}{dx^2} + \frac{d^2 f_1(x)}{dx^2} \\ &= -\frac{2\beta_1^0 \beta_1 B}{N_1 s_1}\left[\left(\frac{\alpha_1^0}{\beta_1^0} - \frac{\alpha_1}{\beta_1 s_1}\right)\frac{1}{\sqrt{y_1(x)y_1(x)}}\right] \\ &\quad - \frac{2\beta_2^0 \beta_2 B}{N_2 s_2}\left[\left(\frac{\alpha_2^0}{\beta_2^0} - \frac{\alpha_2}{\beta_2 s_2}\right)\frac{1}{\sqrt{y_2(x)y_2(x)}}\right] < 0.\end{aligned}$$

Here, the strict inequality follows from the hypothesis of the lemma, and because both $y_1(x)$ and $y_2(x)$ are positive. This proves that the objective is strictly concave. □

Interestingly, the hypothesis of this lemma requires that the ratio of the dose-response parameters alpha and beta for the tumor be bigger than the corresponding ratio for the OAR scaled by the sparing factor, for each modality. Recall that this is the same condition we had encountered in Chapters 2 and 3. It is expected to be met for cancers such as head-and-neck and lung. Algorithm 7 did not incorporate this lemma. We leave that to the reader in one of the exercises below.

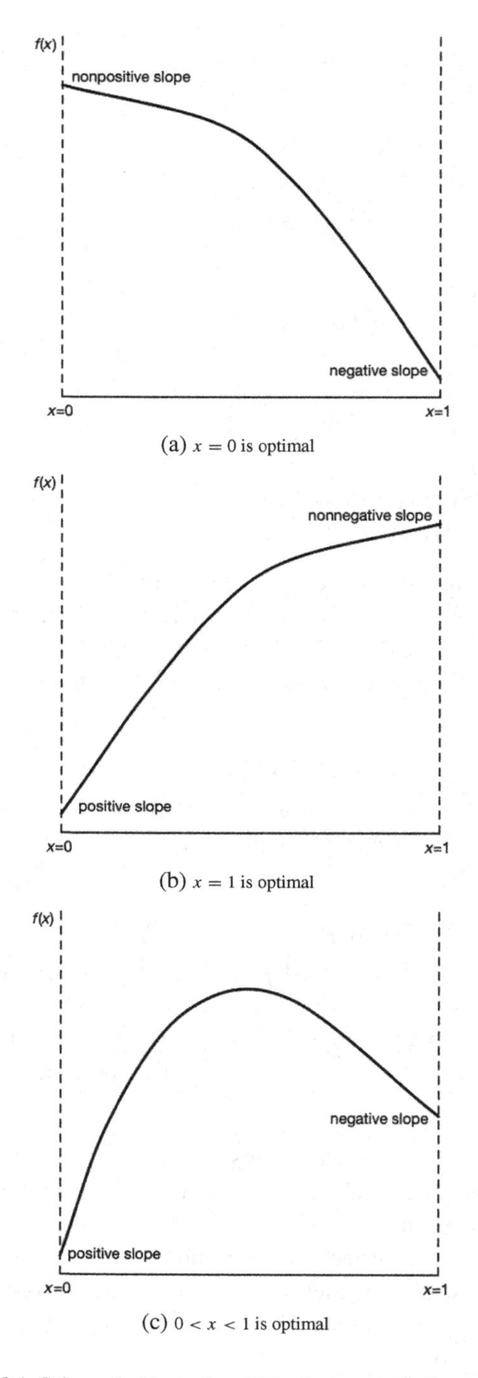

(a) $x = 0$ is optimal

(b) $x = 1$ is optimal

(c) $0 < x < 1$ is optimal

Figure 8.1 Schematic illustration of the three cases in Lemma 8.1.

8.3 Extension to Multiple OAR

The methodology in the above two sections can be extended to tackle problem (2Q(N_1, N_2)), which includes $M \geq 1$ OAR, as follows. At least one of the M OAR constraints (8.2) must be active at an optimal solution. As such, there are two possibilities.

If exactly one constraint is active at an optimal solution, then all candidate d_1, d_2 pairs can be obtained by utilizing the above analyses of the single OAR case (from either Section 8.1 or Section 8.2). This can be done by ignoring all OAR constraints except one, forcing that one OAR constraint to be active, and then keeping only those candidate d_1, d_2 pairs that are feasible for the OAR constraints that were omitted.

If at least two OAR constraints are active, then there are no remaining degrees of freedom in the problem. That is because at most two active constraints are needed to identify a feasible d_1, d_2 pair. Specifically, any two active OAR constraints yield a system of two quadratic equations in variables d_1, d_2. We can solve this system for nonnegative d_1, d_2 and keep any resulting real-valued solution if it is feasible to the $M - 2$ constraints that were not utilized in deriving this d_1, d_2 pair. Observe that there are $\binom{M}{2}$ distinct ways to render two of the M OAR constraints active. Thus, an optimal solution to (2Q(N_1, N_2)) can be determined by comparing the objective values of all (finitely many) candidate pairs discovered through this two-part process. Thus, it only remains to discuss a procedure to solve a system of two equations of the form

$$a_1 x + b_1 x^2 + p_1 y + q_1 y^2 = r_1 \tag{8.41}$$

$$a_2 x + b_2 x^2 + p_2 y + q_2 y^2 = r_2, \tag{8.42}$$

in nonnegative variables x, y. These two generic nonnegative variables take on the role of doses d_1, d_2, respectively. Similarly, a_1, b_1, p_1, q_1, r_1 and a_2, b_2, p_2, q_2, r_2 are positive constants that take on the role of parameters such as $N_1 \alpha_{1m} s_{1m}$ and $N_1 \beta_{1m} s_{1m}^2$ that appear on the left- and right-hand sides of constraint (8.2).

To solve this system of two equations, we first multiply (8.42) with q_1/q_2 to obtain

$$\frac{q_1}{q_2} a_2 x + \frac{q_1}{q_2} b_2 x^2 + \frac{q_1}{q_2} p_2 y + q_1 y^2 = \frac{q_1}{q_2} r_2. \tag{8.43}$$

We subtract this from (8.41) to eliminate y^2. This yields

$$\left(a_1 - \frac{q_1}{q_2} a_2\right) x + \left(b_1 - \frac{q_1}{q_2} b_2\right) x^2 + \left(p_1 - \frac{q_1}{q_2} p_2\right) y = \left(r_1 - \frac{q_1}{q_2} r_2\right). \tag{8.44}$$

If $p_1 - \frac{q_1}{q_2}p_2$, which is the coefficient of y in this equation, equals 0, then this equation reduces to a quadratic equation in x. It can then be solved. If it has any nonnegative real solution, we can substitute it back into (8.41) to find nonnegative solutions of y (if any exist). If, on the other hand, $p_1 - \frac{q_1}{q_2}p_2 \neq 0$, we can divide (8.44) throughout by $\left(p_1 - \frac{q_1}{q_2}p_2\right)$ and express y as a quadratic expression in x. This yields

$$y = \frac{\left(r_1 - \frac{q_1}{q_2}r_2\right) - \left(a_1 - \frac{q_1}{q_2}a_2\right)x - \left(b_1 - \frac{q_1}{q_2}b_2\right)x^2}{\left(p_1 - \frac{q_1}{q_2}p_2\right)}. \tag{8.45}$$

Substituting this expression for y back into (8.41) gives a quartic equation in x. This is written as

$$a_1x + b_1x^2 + p_1\left\{\frac{\left(r_1 - \frac{q_1}{q_2}r_2\right) - \left(a_1 - \frac{q_1}{q_2}a_2\right)x - \left(b_1 - \frac{q_1}{q_2}b_2\right)x^2}{\left(p_1 - \frac{q_1}{q_2}p_2\right)}\right\}$$
$$+p_2\left\{\frac{\left(r_1 - \frac{q_1}{q_2}r_2\right) - \left(a_1 - \frac{q_1}{q_2}a_2\right)x - \left(b_1 - \frac{q_1}{q_2}b_2\right)x^2}{\left(p_1 - \frac{q_1}{q_2}p_2\right)}\right\}^2 = r_1. \tag{8.46}$$

After some algebraic simplification, this quartic equation can be written in the standard form $k_0x^4 + k_1x^3 + k_2x^2 + k_3x + k_4 = 0$, with appropriate values of coefficients k_0, k_1, k_2, k_3, k_4. We solve this quartic equation for x and substitute each nonnegative real solution back into (8.45) to obtain a real value of y. We keep this value of y if it is nonnegative.

8.4 Numerical Experiments

We present results for the single OAR case with $T_1 = 35$ sessions and $\delta_1 = 1.2857$ Gy for that OAR. Some of the other parameter values were fixed throughout as follows: $\alpha_1^0 = 0.35$ Gy^{-1}; $\beta_1^0 = 0.035$ Gy^{-2}; $\rho_1 = 1/3$ Gy^{-1}; $\alpha_1 = 0.35$ Gy^{-1}; and $s_1 = 1$. As before, $N_{\max}$ was fixed at 100. We repeated all experiments with both Algorithms 6 and 7. The results were identical.

We set $\alpha_2^0 = f\alpha_1^0$, $\beta_2^0 = f\beta_1^0$, $\beta_1 = \rho_1\alpha_1$, $\alpha_2 = f\alpha_1$, and $\beta_2 = f\beta_1$. Here, $f > 0$ is a parameter that incorporates the biological power of the second modality. As such, all dose-response parameters of the second modality are scaled by a factor of f as compared to the corresponding parameters of the first modality. On the one hand, when $f > 1$, the second modality is biologically more powerful, that is, causes more damage than the conventional (first) modality, for the same amount of dose administered. On the other hand, when $f < 1$, the second modality is biologically less powerful than the first. Finally, when $f = 1$, the biological power of the two modalities is identical. We conducted numerical experiments with $f \in \{0.2, 0.4, \ldots, 2\}$.

The sparing factors for the second modality were chosen from the set $s_2 \in \{0.9, 1, 1.1\}$. When $s_2 < 1 = s_1$, the second modality has better physical characteristics than the first. This holds because less dose is administered to the OAR by the second modality than the first modality, for an identical tumor dose. The situation is reversed when $s_2 > 1 = s_1$. Finally, the two modalities are physically identical when $s_2 = s_1 = 1$.

Table 8.1 reports the optimal number of sessions for the two modalities, when $T_{\text{double}} = 3$ days. These are denoted by $N_1^\star, N_2^\star$, respectively. It also lists the corresponding optimal doses per session for the two modalities. These are denoted by $d_1^\star, d_2^\star$. The row marked % shows the percentage improvement in objective value achieved by this optimal plan over an optimal fractionation plan that only uses the conventional (first) modality. Dark gray cells highlight cases where it is optimal to only use the first modality. In light gray cells, it is optimal to only use the second modality. The gray cell highlights the case where $f = 1$ and $s_2 = 1$, which renders the two modalities biologically and physically identical. The treatment planner is indifferent between the two modalities in this case, and any combination of N_1, N_2 such that $N_1 + N_2 = 11$ is optimal with a dose of 2.95 Gy per session.

We first consider the panel in the middle, where $s_2 = 1$ and hence the second modality is physically identical to the first. Thus, when $f = 1$, that is, when the second modality is also biologically identical to the first, the two modalities are entirely indistinguishable. This is highlighted in gray. As expected, the percentage improvement over an optimal fractionation plan that only uses the first modality is 0. As the second modality becomes biologically less powerful toward the left of that cell, where $f = 0.8, 0.6, 0.4, 0.2$, the optimal dosing plan only includes the first modality. Consequently, the percentage improvement is 0, as expected. As the second modality becomes biologically more powerful toward the right, where $f = 1.2, 1.4, 1.6, 1.8, 2.0$, the optimal dosing plan only includes the second modality. The percentage improvement increases with the biological power of the second modality.

Table 8.1. *Optimal number of sessions and corresponding optimal doses per session for the two modalities, for various values of f and s_2. The row marked % shows the percentage improvement in objective value over an optimal plan that only uses the first modality.*

	f									
$s_2 = 0.9$	0.2	0.4	0.6	0.8	1.0	1.2	1.4	1.6	1.8	2.0
$N_1^\star$	11	11	0	0	0	0	0	0	0	0
$N_2^\star$	0	0	11	12	12	12	12	12	12	12
$d_1^\star$	2.95	2.95	0.00	0.00	0.00	0.00	0.00	0.00	0.00	0.00
$d_2^\star$	0.00	0.00	4.57	3.58	3.09	2.73	2.45	2.23	2.05	1.90
%	0.00	0.00	5.42	11.61	16.62	20.78	24.34	27.42	30.13	32.54

	f									
$s_2 = 1$	0.2	0.4	0.6	0.8	1.0	1.2	1.4	1.6	1.8	2.0
$N_1^\star$	11	11	11	11	0	0	0	0	0	0
$N_2^\star$	0	0	0	0	11	12	12	12	12	12
$d_1^\star$	2.95	2.95	2.95	2.95	0.00	0.00	0.00	0.00	0.00	0.00
$d_2^\star$	0.00	0.00	0.00	0.00	2.95	2.46	2.21	2.01	1.84	1.71
%	0.00	0.00	0.00	0.00	0.00	3.93	7.28	10.20	12.76	15.04

	f									
$s_2 = 1.1$	0.2	0.4	0.6	0.8	1.0	1.2	1.4	1.6	1.8	2.0
$N_1^\star$	11	11	11	11	11	11	11	11	11	0
$N_2^\star$	0	0	0	0	0	0	0	0	0	11
$d_1^\star$	2.95	2.95	2.95	2.95	2.95	2.95	2.95	2.95	2.95	0.00
$d_2^\star$	0.00	0.00	0.00	0.00	0.00	0.00	0.00	0.00	0.00	1.65
%	0.00	0.00	0.00	0.00	0.00	0.00	0.00	0.00	0.00	1.17

We now consider the top panel, where $s_2 = 0.9$ and hence the second modality is physically superior to the first. Thus, when $f = 1$, that is, when the second modality is biologically identical to the first, it is optimal to only use the second modality. Then, the second modality remains optimal as it becomes biologically more powerful toward the right, where $f = 1.2, 1.4, 1.6, 1.8, 2.0$. The percentage improvement also increases with this enhanced biological power. The second modality also remains optimal in this case when its biological power drops to $f = 0.8, 0.6$, with a corresponding decrease in the percentage improvement. The biological power eventually becomes too weak when $f = 0.4$, where it is optimal to switch to the first modality. In particular, the first modality remains optimal when $f = 0.2$. As expected, the percentage improvement is 0 in these two cases. This is intuitive, because the second modality is physically superior in the top panel, thus a weaker biological power is sufficient for the treatment planner to prefer it, as compared to the middle panel. The percentage improvement values in the top panel dominate the corresponding values in the middle panel for this same reason.

In the bottom panel, the second modality is physically inferior to the first, because $s_2 = 1.1$. Thus, when $f = 1$, that is, when the second modality is biologically identical to the first, it is optimal to only use the first modality. The percentage improvement is thus 0, as expected. The first modality remains optimal as the second modality becomes biologically weaker toward the left, for $f = 0.8, 0.6, 0.4, 0.2$. The percentage improvement is therefore 0 for these values of f. The first modality remains optimal even when the biological power of the second modality increases toward the right, for $f = 1.2, 1.4, 1.6, 1.8$. The percentage improvement therefore remains 0. The biological power of the second modality eventually becomes high enough, rendering it optimal, when $f = 2.0$. The percentage improvement also becomes positive there. As such, the switch from the first modality to the second modality occurs at the lowest value of f in the top panel, the highest value of f in the bottom panel, and a medium value of f in the middle panel. This is intuitive, because the second modality is physically superior in the top panel, thus a weaker biological power is sufficient for the treatment planner to prefer it, as compared to the middle panel. This trend continues to the bottom panel. The percentage improvement values in the upper panels dominate the corresponding values in the lower panels for this same reason.

These experiments were repeated with $T_{\text{double}} = 6$ days. The results are summarized in Table 8.2. All qualitative trends in Table 8.2 are similar to those in Table 8.1. The optimal number of treatment sessions is higher in Table 8.2 than Table 8.1, as expected, because of the slower growing tumor.

Table 8.2. *Results of experiments similar to those reported in Table 8.1, except that the tumor doubling time was fixed at* T_{double} *= 6 days here.*

					f					
$s_2 = 0.9$	0.2	0.4	0.6	0.8	1.0	1.2	1.4	1.6	1.8	2.0
$N_1^\star$	23	23	0	0	0	0	0	0	0	0
$N_2^\star$	0	0	25	25	24	24	24	23	23	22
$d_1^\star$	1.76	1.76	0.00	0.00	0.00	0.00	0.00	0.00	0.00	0.00
$d_2^\star$	0.00	0.00	2.65	2.17	1.90	1.66	1.47	1.37	1.25	1.19
%	0.00	0.00	5.03	10.84	15.33	18.94	21.94	24.48	26.68	28.61

					f					
$s_2 = 1$	0.2	0.4	0.6	0.8	1.0	1.2	1.4	1.6	1.8	2.0
$N_1^\star$	23	23	23	23	0	0	0	0	0	0
$N_2^\star$	0	0	0	0	23	23	22	22	22	21
$d_1^\star$	1.76	1.76	1.76	1.76	0.00	0.00	0.00	0.00	0.00	0.00
$d_2^\star$	0.00	0.00	0.00	0.00	1.76	1.54	1.42	1.28	1.17	1.12
%	0.00	0.00	0.00	0.00	0.00	3.43	6.28	8.71	10.80	12.65

					f					
$s_2 = 1.1$	0.2	0.4	0.6	0.8	1.0	1.2	1.4	1.6	1.8	2.0
$N_1^\star$	23	23	23	23	23	23	23	23	23	23
$N_2^\star$	0	0	0	0	0	0	0	0	0	0
$d_1^\star$	1.76	1.76	1.76	1.76	1.76	1.76	1.76	1.76	1.76	1.76
$d_2^\star$	0.00	0.00	0.00	0.00	0.00	0.00	0.00	0.00	0.00	0.00
%	0.00	0.00	0.00	0.00	0.00	0.00	0.00	0.00	0.00	0.00

Bibliographic Notes

Algebraic methods for solving quartic equations are discussed in [12, 90]. The formulation in this chapter, the solution method in Section 8.1, and its extension to the multiple OAR case in Section 8.3 are based on the doctoral dissertation [91] and the associated paper [93]. We refer the reader to the extensive numerical experiments there for further insights. A brief summary of this work was included in a tutorial [61], without listing Algorithm 6 and the accompanying numerical results. The solution method in Section 8.2 is new.

Exercises

Exercise 8.2 Express the quartic equation (8.46) in the standard form $k_0x^4 + k_1x^3 + k_2x^2 + k_3x + k_4 = 0$ by deriving appropriate values of coefficients k_0, k_1, k_2, k_3, k_4. Incorporate this into a computer program that can tackle the multiple OAR case using (i) the KKT approach from Section 8.1, and then (ii) the single-variable transformation from Section 8.2.

Exercise 8.3 Design numerical experiments and perform sensitivity analyses using your computer program from the above exercise. Discuss any qualitative trends you observe.

Exercise 8.4 Explore how you would incorporate the insight from Lemma 8.1 into Algorithm 7.

Exercise 8.5 *The numerical experiments in this chapter did not reveal a scenario where a combination of modalities was strictly better than using any single modality. That is, a scenario where $N_1^{\text{optimal}} > 0$, $d_1^{\text{optimal}} > 0$, $N_2^{\text{optimal}} > 0$, $d_2^{\text{optimal}} > 0$ and the corresponding optimal objective value was strictly better than any solution where at least one of these four quantities was 0. Either prove that such a scenario cannot occur or find parameter values where it does occur. As a related matter, either prove that the third possibility in Lemma 8.1 can never occur or find parameter values where it does.

9

Robust Fractionation with Two Modalities

We now extend the formulation in the previous chapter to incorporate uncertainty in problem parameters via a robust approach. This extension is thus akin to our efforts in Chapter 4 relative to Chapter 3.

The treatment planner assumes that the sparing factors s_i and OAR dose-response parameters α^i_m, β^i_m belong to intervals of positive real numbers, for $i = 1,2$. We denote these by $[s^{\min}_{im}, s^{\max}_{im}]$, $[\alpha^{\min}_{im}, \alpha^{\max}_{im}]$, and $[\beta^{\min}_{im}, \beta^{\max}_{im}]$, for $i = 1,2$ and $m \in \mathcal{M}$. For instance, if the second modality is protons, the interval for sparing factor s_2 could model uncertainty in the location of the Bragg peak described in Chapter 1. The treatment planner wishes to compute a dosing plan that attains the largest biological effect on the tumor, while abiding by the OAR constraints regardless of the actual value of the uncertain parameters. We thus consider the robust counterpart

$$(2QR) \quad \max_{N_1,d_1,N_2,d_2} \sum_{i=1,2} N_i \left[\alpha^0_i d_i + \beta^0_i (d_i)^2\right] - \tau(N_1 + N_2)$$

$$\sum_{i=1,2} N_i[\alpha_{im}(s_{im}d_i) + \beta_{im}(s_{im}d_i)^2] \le T_{1m}\delta_{1m}(\alpha_{1m} + \beta_{1m}\delta_m),$$

$$\forall s_{im} \in [s^{\min}_{im}, s^{\max}_{im}],\ \forall \alpha_{im} \in [\alpha^{\min}_{im}, \alpha^{\max}_{im}],\ \forall \beta_{im} \in [\beta^{\min}_{im}, \beta^{\max}_{im}],$$

$$i = 1,2,\ m \in \mathcal{M} \tag{9.1}$$

$$d_1 \ge 0,\ d_2 \ge 0 \tag{9.2}$$

$$0 \le N_1, \text{ integer};\ 0 \le N_2, \text{ integer} \tag{9.3}$$

$$1 \le N_1 + N_2 \le N_{\max}, \tag{9.4}$$

of the nominal formulation (2Q) from Chapter 8.

Similar to previous chapters, we solve this problem for each fixed pair of nonnegative integers (N_1, N_2) such that $1 \le N_1 + N_2 \le N_{\max}$. We denote each of these problems by $(2QR(N_1, N_2))$. We choose a feasible pair (N_1, N_2) and the corresponding optimal dosing pair (d_1, d_2) that attain the largest objective value over all problems (2QR(N_1, N_2)).

9.1 Reformulation as a Finite Set of Subproblems

A pair of doses $(d_1 \geq 0, d_2 \geq 0)$ satisfies constraint (9.1) for OAR $m \in \mathcal{M}$ in problem (2QR(N_1, N_2)) if, and only if, the pair satisfies constraint (8.2) in Chapter 8 when $s_{im} = s_{im}^{\max}$, for $i = 1, 2$; $\alpha_{2m} = \alpha_{2m}^{\max}$, and $\beta_{2m} = \beta_{2m}^{\max}$. We call these the bottleneck values of these parameters. Parameters α_{1m}, β_{1m} need a more careful analysis because they in general appear on both sides of constraint (9.1).

We first tackle the special case $N_1 = 0$ whereby α_{1m}, β_{1m} only appear on the right-hand side. Thus, their bottleneck values are $\alpha_{1m}^{\min}, \beta_{1m}^{\min}$. This yields the candidate dosing plan $d_1 = 0$ and

$$d_2 = \min_{m \in \mathcal{M}} \left\{ \frac{-\alpha_{2m}^{\max} + \sqrt{(\alpha_{2m}^{\max})^2 + 4\beta_{2m}^{\max}(T_m \delta_m(\alpha_{1m}^{\min} + \beta_{1m}^{\min}\delta_m))/N_2}}{2\beta_{2m}^{\max} s_{2m}^{\max}} \right\}. \tag{9.5}$$

Here, since $N_1 = 0$, the value of d_1 does not matter. It was thus set to 0 arbitrarily, and then d_2 was derived as in formula (3.9) from Chapter 3. Observe that when $N_1 = 0$, the constraint $N_1 + N_2 \geq 1$ ensures that $N_2 \geq 1$. Thus, the division by N_2 in formula (9.5) does not cause any issues.

Now, when $N_1 \neq 0$, we rewrite the constraint by collecting α_{1m} and β_{1m} terms on one side as

$$\begin{aligned} &\alpha_{1m}[N_1(s_{1m}^{\max} d_1) - T_m \delta_m] + \beta_{1m}[N_1(s_{1m}^{\max} d_1)^2 - T_m(\delta_m)^2] \\ &\quad + N_2[\alpha_{2m}^{\max}(s_{2m}^{\max} d_2) + \beta_{2m}^{\max}(s_{2m}^{\max} d_2)^2] \leq 0, \ \forall \alpha_{1m} \in [\alpha_{1m}^{\min}, \alpha_{1m}^{\max}], \\ &\quad \forall \beta_{1m} \in [\beta_{1m}^{\min}, \beta_{1m}^{\max}]. \end{aligned} \tag{9.6}$$

We make four observations about constraint (9.6).

O1. If $d_1 \geq \frac{T_m \delta_m}{N_1 s_{1m}^{\max}}$, then the bottleneck value of α_{1m} is $\alpha_{1m}^{\max}$.
O2. If $d_1 \leq \frac{T_m \delta_m}{N_1 s_{1m}^{\max}}$, then the bottleneck value of α_{1m} is $\alpha_{1m}^{\min}$.
O3. If $(d_1)^2 \geq \frac{T_m(\delta_m)^2}{N_1(s_{1m}^{\max})^2}$, then the bottleneck value of β_{1m} is $\beta_{1m}^{\max}$.
O4. If $(d_1)^2 \leq \frac{T_m(\delta_m)^2}{N_1(s_{1m}^{\max})^2}$, then the bottleneck value of β_{1m} is $\beta_{1m}^{\min}$.

We thus sort the OAR in $\mathcal{M}$ according to increasing values of quantities $\mu_m = \frac{T_m \delta_m}{N_1 s_{1m}^{\max}}$. We store this sorted list as the ordered tuple $\mathcal{I} = (I_1, \ldots, I_M)$.

That is, $\mu_{I_k} \leq \mu_{I_{k+1}}$, for $k = 1, \ldots, M-1$. We also sort the OAR in $\mathcal{M}$ according to increasing values of quantities $\nu_m = \frac{T_m(\delta_m)^2}{N_1(s_{1m}^{\max})^2}$. We store this sorted list as another ordered tuple $\mathcal{J} = (J_1, \ldots, J_M)$. That is, $\nu_{J_\ell} \leq \nu_{J_{\ell+1}}$, for $\ell = 1, \ldots, M-1$. This sorting allows us to decompose the set of all $d_1 \geq 0, (d_1)^2$ pairs into subregions indexed by (k, ℓ), for $k = 0, 1, \ldots, M$ and $\ell = 0, 1, \ldots, M$. Subregion (k, ℓ) includes the constraints $\mu_{I_k} \leq d_1 \leq \mu_{I_{k+1}}$ and $\nu_{J_\ell} \leq (d_1)^2 \leq \nu_{J_{\ell+1}}$. Here, we interpret $I_0 = J_0 = 0$ and $I_{M+1} = J_{M+1} = M+1$. Moreover, we also interpret $\mu_{I_0} = \nu_{J_0} = 0$, and $\mu_{I_{M+1}} = \nu_{J_{M+1}} = \infty$. Equivalently, it includes constraints

$$\underbrace{\max\{\mu_{I_k}, \sqrt{\nu_{J_\ell}}\}}_{\theta_{k,\ell}} \leq d_1 \leq \underbrace{\min\{\mu_{I_{k+1}}, \sqrt{\nu_{J_{\ell+1}}}\}}_{\Omega_{k,\ell}}. \tag{9.7}$$

We leave it to the readers to verify that $\theta_{00} = 0$; $\theta_{0\ell} > 0$, for $\ell \neq 0$; $\theta_{k0} > 0$, for $k \neq 0$; $\Omega_{M,M} = \infty$; $\Omega_{M,\ell} < \infty$, for $\ell \neq M$; and $\Omega_{k,M} < \infty$, for $k \neq M$.

We interpret sets $\{I_1, \ldots, I_0\} = \{I_1, \ldots, 0\}$; $\{J_1, \ldots, J_0\} = \{J_1, \ldots, 0\}$; $\{I_{M+1}, \ldots, I_M\} = \{M+1, \ldots, I_M\}$; and $\{J_{M+1}, \ldots, J_M\} = \{M+1, \ldots, J_M\}$ as empty sets. Then, in subregion (k, ℓ), the bottleneck values of α_{1m}, β_{1m} are given by

$$\alpha_{1m}^{\max}, \beta_m^{1\max}, \text{ for } m \in \{I_1, \ldots, I_k\} \cap \{J_1, \ldots, J_\ell\}, \tag{9.8}$$

$$\alpha_{1m}^{\max}, \beta_m^{1\min}, \text{ for } m \in \{I_1, \ldots, I_k\} \cap \{J_{\ell+1}, \ldots, J_M\}, \tag{9.9}$$

$$\alpha_{1m}^{\min}, \beta_m^{1\max}, \text{ for } m \in \{I_{k+1}, \ldots, I_M\} \cap \{J_1, \ldots, J_\ell\}, \tag{9.10}$$

$$\alpha_{1m}^{\min}, \beta_m^{1\min}, \text{ for } m \in \{I_{k+1}, \ldots, I_M\} \cap \{J_{\ell+1}, \ldots, J_M\}. \tag{9.11}$$

The logic behind statements (9.8)–(9.11) is explained below.

Consider (9.8). Since OAR $m \in \{I_1, \ldots, I_k\} \cap \{J_1, \ldots, J_\ell\}$, we know that $\mu_m \leq \mu_{I_k} \leq \max\{\mu_{I_k}, \sqrt{\nu_{J_\ell}}\}$ and $\sqrt{\nu_m} \leq \sqrt{\nu_{J_\ell}} \leq \max\{\mu_{I_k}, \sqrt{\nu_{J_\ell}}\}$. The first inequality in the defining constraints (9.7) for subregion (k, ℓ) thus implies that $\mu_m \leq d_1$ and $\sqrt{\nu_m} \leq d_1$. The definitions of μ_m, ν_m and observations O1, O3 above then imply (9.8).

Now consider (9.9). Since OAR $m \in \{I_1, \ldots, I_k\} \cap \{J_{\ell+1}, \ldots, J_M\}$, we know that $\mu_m \leq \mu_{I_k} \leq \max\{\mu_{I_k}, \sqrt{\nu_{J_\ell}}\}$ and $\sqrt{\nu_m} \geq \sqrt{\nu_{J_{\ell+1}}} \geq \min\{\mu_{I_{k+1}}, \sqrt{\nu_{J_{\ell+1}}}\}$. The first inequality in the defining constraints (9.7) for subregion (k, ℓ) thus implies that $\mu_m \leq d_1$. The second inequality in (9.7) implies that $\sqrt{\nu_m} \geq d_1$. The definitions of μ_m, ν_m and observations O1, O4 above then imply (9.8).

Now consider (9.10). Since OAR $m \in \{I_{k+1}, \ldots, I_M\} \cap \{J_1, \ldots, J_\ell\}$, we know that $\mu_m \geq \mu_{I_{k+1}} \geq \min\{\mu_{I_{k+1}}, \sqrt{\nu_{J_{\ell+1}}}\}$ and $\sqrt{\nu_m} \leq \sqrt{\nu_{J_\ell}} \leq \max\{\mu_{I_k}, \sqrt{\nu_{J_\ell}}\}$. The first inequality in the defining constraints (9.7) for subregion (k,ℓ) thus implies that $\sqrt{\nu_m} \le d_1$. The second inequality in (9.7) implies that $\mu_m \ge d_1$. The definitions of μ_m, ν_m and observations O2, O3 above then imply (9.10).

Finally, consider (9.11). We know that

$$\mu_m \ge \mu_{I_{k+1}} \ge \min\{\mu_{I_{k+1}}, \sqrt{\nu_{J_{\ell+1}}}\} \text{ and}$$
$$\sqrt{\nu_m} \ge \sqrt{\nu_{J_{\ell+1}}} \ge \min\{\mu_{I_{k+1}}, \sqrt{\nu_{J_{\ell+1}}}\},$$

because OAR $m \in \{I_{k+1}, \dots, I_M\} \cap \{J_{\ell+1}, \dots, J_M\}$. The second inequality in the defining constraints (9.7) for subregion (k,ℓ) thus implies that $\mu_m \ge d_1$ and $\sqrt{\nu_m} \ge d_1$. The definitions of μ_m, ν_m and observations O2, O4 above then imply (9.11).

For brevity, we denote the bottleneck values defined in (9.8)–(9.11) simply by $\alpha^\star_{1m}(k,\ell), \beta^\star_{1m}(k,\ell)$. This allows us to reformulate the above robust counterpart as an equivalent collection of subproblems

$$(2\mathrm{QR}_{k,\ell}(N_1,N_2)) \max_{d_1,d_2} \sum_{i=1,2} N_i \left[\alpha_i^0 d_i + \beta_i^0 (d_i)^2\right] - \tau(N_1+N_2)$$

$$N_1[\alpha^\star_{1m}(k,\ell)(s^{\max}_{1m} d_1) + \beta^\star_{1m}(k,\ell)(s^{\max}_{1m} d_1)^2]$$
$$+ N_2[\alpha^{\max}_{2m}(s^{\max}_{2m} d_2) + \beta^{\max}_{2m}(s^{\max}_{2m} d_2)^2]$$
$$\le \underbrace{T_m \delta_m(\alpha^\star_{1m}(k,\ell) + \beta^\star_{1m}(k,\ell)\delta_m)}_{B^\star_m(k,\ell)},\ m \in \mathcal{M} \tag{9.12}$$

subject to (9.7)

$d_1 \ge 0, d_2 \ge 0,$

for $k = 0,1,\dots,M$ and $\ell = 0,1,\dots,M$. The total number of subproblems is $(M+1)^2$. A subproblem with the largest objective function value thus supplies an optimal dosing plan for problem $(2\mathrm{QR}(N_1,N_2))$.

9.2 Solution of Subproblems via a Single Variable Transformation

Problem $(2\mathrm{QR}_{k,\ell}(N_1,N_2))$ has the same form as the nominal problem $(2\mathrm{Q1}(N_1,N_2))$ in Chapter 8, except for the bound constraints in (9.7). Here, we describe how to adapt the single-variable transformation approach from Section 8.2 to problem $(2\mathrm{QR}_{k,\ell}(N_1,N_2))$. For expository simplicity, we explain this in the context of a problem with a single OAR. We leave the extension to the case of multiple OAR to the readers in one of the exercises below.

We reproduce for easy reference this special case of ($2\text{QR}_{k,\ell}(N_1,N_2)$) with a single OAR. We write it as

$$(2\text{QR}1_{k,\ell}(N_1,N_2)) \max_{d_1,d_2} \sum_{i=1,2} N_i\left[\alpha_i^0 d_i + \beta_i^0 (d_i)^2\right] - \tau(N_1+N_2)$$

$$N_1[\alpha_1^\star(k,\ell)(s_1^{\max}d_1) + \beta_1^\star(k,\ell)(s_1^{\max}d_1)^2]$$
$$+ N_2[\alpha_2^{\max}(s_2^{\max}d_2) + \beta_2^{\max}(s_2^{\max}d_2)^2]$$
$$\leq \underbrace{T_1\delta_1(\alpha_1^\star(k,\ell) + \beta_1^\star(k,\ell)\delta_1)}_{B^\star(k,\ell)} \tag{9.13}$$

subject to (9.7)

$d_1 \geq 0, d_2 \geq 0,$

for $k = 0,1$ and $\ell = 0,1$. Note that we have dropped the OAR subscript m from this formulation, as we only have a single OAR.

Since the objective function is increasing in both d_1 and d_2, there are only three possibilities with regard to the pair of constraints (9.7) and (9.13): either (i) the lower bound in (9.7) is active, or (ii) the upper bound in (9.7) is active, or (iii) neither the lower bound nor the upper bound in (9.7) is active but (9.13) is active. Under the first two possibilities, there are two further subcases: (a) there is no $d_2 \geq 0$ that can render (9.13) feasible, or (b) there is a $d_2 \geq 0$ that renders (9.13) active. Note that the largest value of d_1 that constraint (9.13) admits can be calculated by setting $d_2 = 0$ and solving a quadratic equation that renders the constraint active. This value is

$$\bar{d}_1(k,\ell) = \frac{-\alpha_1^\star(k,\ell) + \sqrt{(\alpha_1^\star(k,\ell))^2 + \frac{4\beta_1^\star(k,\ell)B^\star(k,\ell)}{N_1}}}{2\beta_1^\star(k,\ell)s_1^{\max}}. \tag{9.14}$$

Consequently, Case (i)(a) could arise if $\theta_{k,\ell} > \bar{d}_1(k,\ell)$ and Case (ii)(a) could arise if $\Omega_{k,\ell} > \bar{d}_1(k,\ell)$. On the contrary, Case (i)(b) could arise if $\theta_{k,\ell} \leq \bar{d}_1(k,\ell)$ and then we obtain

$$d_2(k,\ell) = \begin{cases} 0, \text{ if } N_2 = 0 \\ \frac{-\alpha_2^{\max} + \sqrt{(\alpha_2^{\max})^2 + \frac{\Delta_2(\theta_{k,\ell})}{N_2}}}{2\beta_2^{\max}s_2^{\max}}, \text{ if } N_2 \neq 0, \end{cases} \tag{9.15}$$

as long as $\theta_{k,\ell} \leq \Omega_{k,\ell}$ so that setting $d_1 = \theta_{k,\ell}$ abides by constraint (9.7). Here, we have employed the shorthand

$$\Delta_2(\theta_{k,\ell}) = 4\beta_2^{\max}(B^\star(k,\ell) - N_1[\alpha_1^\star(k,\ell)(s_1^{\max}\theta_{k\ell}) + \beta_1^\star(k,\ell)(s_1^{\max}\theta_{k,\ell})^2]).$$

Similarly, Case (ii)(b) could arise if $\Omega_{k,\ell} \leq \bar{d}_1(k,\ell)$ and then we obtain

$$d_2(k,\ell) = \begin{cases} 0, \text{ if } N_2 = 0 \\ \dfrac{-\alpha_2^{\max} + \sqrt{(\alpha_2^{\max})^2 + \frac{\Delta_2(\Omega_{k,\ell})}{N_2}}}{2\beta_2^{\max} s_2^{\max}}, \text{ if } N_2 \neq 0, \end{cases} \tag{9.16}$$

as long as $\theta_{k,\ell} \leq \Omega_{k,\ell}$ so that setting $d_1 = \Omega_{k,\ell}$ abides by constraint (9.7). Since neither the lower bound nor the upper bound in (9.7) is binding under Case (iii), the situation is similar to Chapter 8. Thus, if $N_2 = 0$, we let $d_2 = 0$ and $d_1 = \bar{d}_1(k,\ell)$, as long as $\theta_{k,\ell} < \bar{d}_1(k,\ell) < \Omega_{k,\ell}$. If $N_2 \neq 0$, we introduce a variable $0 \leq x \leq 1$ to denote the fraction of $B^\star(k,\ell)$ that is administered using the first modality, similar in principle to Chapter 8. Thus, all possible candidate values of $0 \leq x \leq 1$ can be obtained via calculations identical to Section 8.2 from that chapter. The corresponding values of doses are given by

$$d_1(x;k,\ell) = \frac{-\alpha_1^\star(k,\ell) + \sqrt{(\alpha_1^\star(k,\ell))^2 + \frac{4\beta_1^\star(k,\ell)xB^\star(k,\ell)}{N_1}}}{2\beta_1^\star(k,\ell)s_1^{\max}}, \text{ and} \tag{9.17}$$

$$d_2(x;k,\ell) = \frac{-\alpha_2^{\max} + \sqrt{(\alpha_2^{\max})^2 + \frac{4\beta_2^{\max}(1-x)B^\star(k,\ell)}{N_2}}}{2\beta_2^{\max} s_2^{\max}}. \tag{9.18}$$

This pair provides candidate optimal doses if $\theta_{k,\ell} < d_1(x;k,\ell) < \Omega_{k,\ell}$, as required by Case (iii). The overall procedure is listed in Algorithm 8.

*Use $\alpha^0, \beta^0, \alpha_1^\star, \beta_1^\star, s_1^{\max}, \alpha_2^{\max}, \beta_2^{\max}, s_2^{\max}, B^\star, N_1, N_2, \theta$, and Ω as input. In Steps 4, 6, 11, 16, and 23 of Algorithm 7, admit (d_1, d_2) as candidates only if $\theta < d_1 < \Omega$.

9.3 Numerical Experiments

In this section, we present results obtained by applying Algorithm 8 to an example with a single OAR. For this OAR, we use $T_1 = 35$ sessions and $\delta_1 = 1.2857$ Gy as before. We fix throughout $N_{\max} = 100$; $\alpha_1^0 = 0.35$ Gy^{-1}; $\beta_1^0 = 0.035$ Gy^{-2}; $\alpha_1 = 0.35$ Gy^{-1}; $\beta_1 = \alpha_1/3$ Gy^{-2} (that is, $\rho_1 = 1/3$ Gy^{-1}); and $s_1^{\min} = s_1^{\max} = 1$. As such, we do not include any uncertainty in the sparing factor for the conventional (first) modality. We discuss two sets of experiments. In both sets, we study the effect of uncertainty in the OAR dose-response parameters of the two modalities.

In the first set of experiments (Section 9.3.1), the two modalities have identical (nominal) dose-response parameters but the sparing factor of the second modality is smaller than the first. In this sense, the second modality is

Algorithm 8 Exact solution of (2QR) with a single OAR

1: Input - α_i^0, β_i^0, for $i = 1, 2$, and T_{double} for the tumor; $N_{\max}$.
2: Input - T, δ; and $\alpha_i^{\min}, \beta_i^{\max}, s_i^{\min}, s_i^{\max}$, for $i = 1, 2$, for the OAR.
3: Let $I_0 = J_0 = 0$; $I_1 = J_1 = 1$; and $I_2 = J_2 = 2$. ▷ Sorting not necessary.
4: Let $\mu_0 = \nu_0 = 0$ and $\mu_2 = \nu_2 = \infty$.
5: best $= -\infty$.
6: **for** $N_1 = 0 : N_{\max}$ **do**
7: **for** $N_2 = \max\{0, 1 - N_1\} : N_{\max} - N_1$ **do**
8: Let $\mathcal{L} = \{\}$.
9: **if** $N_1 = 0$ **then** ▷ It is guaranteed that $N_2 \neq 0$.
10: Let $d_1 = 0$ and d_2 as in (9.5). Append (d_1, d_2) to $\mathcal{L}$.
11: **else** ▷ $N_1 \neq 0$.
12: Calculate $\mu_1 = \frac{T\delta}{N_1 s_1^{\max}}$ and $\nu_1 = \frac{T\delta^2}{N_1 (s_1^{\max})^2}$.
13: **for** $k = 0, 1$ **do**
14: **for** $\ell = 0, 1$ **do**
15: subprob$\mathcal{L} = \{\}$.
16: $\alpha_1^\star(k, \ell), \beta_1^\star(k, \ell)$ per (9.8)-(9.11), $m = 1$.
17: Calculate $B^\star = T\delta(\alpha_1^\star(k, \ell) + \beta_1^\star(k, \ell)\delta)$.
18: Calculate $\bar{d}_1$ via (9.14).
19: Let $\theta = \max\{\mu_k, \sqrt{\nu_\ell}\}$ and $\Omega = \min\{\mu_{k+1}, \sqrt{\nu_{\ell+1}}\}$.
20: **if** $\theta \leq \bar{d}_1$ and $\theta \leq \Omega$ **then** ▷ Case (i)(b).
21: Let $d_1 = \theta$; d_2 via (9.15).
Append (d_1, d_2) to subprob$\mathcal{L}$.
22: **end if**
23: **if** $\Omega \leq \bar{d}_1$ and $\theta \leq \Omega$ **then** ▷ Case (ii)(b).
24: Let $d_1 = \Omega$; d_2 via (9.16).
Append (d_1, d_2) to subprob$\mathcal{L}$.
25: **end if**
26: **if** $N_2 = 0$ **then**
27: **if** $\theta < \bar{d}_1 < \Omega$ **then,**
28: Let $d_1 = \bar{d}_1$, $d_2 = 0$.
Append (d_1, d_2) to subprob$\mathcal{L}$.
29: **end if**
30: **else** ▷ $N_2 \neq 0$.
31: Append doses to subprob$\mathcal{L}$ (Steps* 2-26 Alg. 7).
32: **end if**
33: Find doses in subprob$\mathcal{L}$ with largest obj.
34: **end for** ▷ Loop over ℓ.
35: **end for** ▷ Loop over k.
36: Find a subprob. with largest obj.; append optimal doses to $\mathcal{L}$.
37: **end if** ▷ Checking whether or not $N_1 = 0$.
38: Steps 20–26 from Algorithm 6.
39: **end for** ▷ Loop over N_2.
40: **end for** ▷ Loop over N_1.
41: Output - $N_1^{\text{optimal}}, N_2^{\text{optimal}}, d_1^{\text{optimal}}, d_2^{\text{optimal}}$.

biologically similar to the first, but it has a superior dose-deposition profile. In the second set of experiments (Section 9.3.2), the two modalities have identical sparing factors but the (nominal) dose-response parameters of the second modality are bigger than the first. In this sense, the second modality has a dose-deposition profile that is similar to the first, but the second modality is biologically more powerful.

9.3.1 Alternative Modality with Superior Dose-Deposition Profile

In this section, we focus on the scenario where the nominal $\alpha_2^0 = \alpha_1^0$; $\beta_2^0 = \beta_1^0$; $\alpha_2 = \alpha_1$; $\beta_2 = \beta_1$. That is, the nominal biological power of the second modality is identical to the first. As an aside, note that the nominal $\rho_2 = \beta_2/\alpha_2 = \beta_1/\alpha_1 = \rho_1$. We will thus denote this ratio for both modalities simply by ρ. We model the uncertainty in the OAR dose-response parameters of both modalities via intervals similar to Chapters 4 and 7. Specifically, we let $\alpha_i^{\min} = \alpha_i(1-\Delta_i)$; $\alpha_i^{\max} = \alpha_i(1+\Delta_i)$; $\beta_i^{\min} = \beta_i(1-\Delta_i)$; $\beta_i^{\max} = \beta_i(1+\Delta_i)$, for $i = 1,2$, where $0 \leq \Delta_i < 1$ are parameters that control the length of the uncertainty intervals for the two modalities. As before, we note that $\Delta_i = 0$ recovers the nominal value of the corresponding dose-response parameter, and the larger the Δ_i, the larger the uncertainty interval. We report results for $\Delta_i \in \{0, 0.1, \ldots, 0.9\}$, for $i = 1,2$. To ensure that the dose-deposition profile of the second modality is superior to the first, we fix its sparing factor at $s_2 = 0.5$ without any uncertainty. The results are displayed in Table 9.1.

The five panels within Table 9.1 display N_1^{optimal}, N_2^{optimal}, d_1^{optimal}, d_2^{optimal}, and the percentage price of robustness, from top to bottom, respectively. In the N_1^{optimal} and N_2^{optimal} panels, the cells where only the first modality is optimal are colored dark gray; the cells where only the second modality is optimal are colored light gray; and the cells where a combination of the two modalities is optimal are colored gray. In the nominal case where $\Delta_1 = \Delta_2 = 0$, both modalities are biologically identical. Thus, since $s_2 = 0.5 < s_1 = 1$, it is optimal to administer only the second modality. Consequently, $N_2^{\text{optimal}} > 0, d_2^{\text{optimal}} > 0$, but $N_1^{\text{optimal}} = 0, d_1^{\text{optimal}} = 0$. In general, whenever $N_1^{\text{optimal}} = 0$ and $d_1^{\text{optimal}} = 0$, we can derive the counterpart

$$N_2^{\star} = \frac{4\rho(T\delta(\alpha_1 + \beta_1\delta)(1-\Delta_1))/(\alpha_2(1+\Delta_2))}{\left(\frac{\eta+\sqrt{\eta^2+2\eta r_2(\alpha_0-\alpha_2\beta_0/(s_2\beta_2))}}{r_2(\alpha_0-\alpha_2\beta_0/(s_2\beta_2))}+1\right)^2-1} \tag{9.19}$$

of formula (2.44) from Chapter 2. Here, $r_2 = 1/(2s_2\rho)$. Formula (9.19) can then be utilized in an appropriate counterpart of Theorem 2.4 to obtain

Table 9.1. *Results for Section 9.3.1 with different levels of uncertainty in the dose-response parameters of the two modalities. Decimals were suppressed in the bottom table labeled % to save space.*

	$N_1^{optimal}$	Δ_2: 0	0.1	0.2	0.3	0.4	0.5	0.6	0.7	0.8	0.9
Δ_1	0	0	0	0	0	0	0	0	0	0	0
	0.1	0	0	0	0	0	0	0	0	0	0
	0.2	0	0	0	0	0	0	0	0	0	0
	0.3	0	0	0	0	0	0	0	0	0	0
	0.4	0	0	0	0	0	0	0	7	8	10
	0.5	0	0	0	0	7	8	10	11	12	12
	0.6	0	0	8	10	11	12	13	14	14	15
	0.7	10	12	13	14	15	16	16	17	18	18
	0.8	16	17	18	19	19	20	21	21	26	26
	0.9	23	29	29	29	29	29	29	29	29	29

	$N_2^{optimal}$	Δ_2: 0	0.1	0.2	0.3	0.4	0.5	0.6	0.7	0.8	0.9
Δ_1	0	13	12	11	10	9	9	8	8	7	7
	0.1	12	11	10	9	9	8	7	7	7	6
	0.2	11	10	9	8	8	7	7	6	6	6
	0.3	9	8	8	7	7	6	6	5	5	5
	0.4	8	7	7	6	6	5	5	2	2	1
	0.5	7	6	6	5	2	2	1	1	1	1
	0.6	5	5	2	1	1	1	1	1	1	1
	0.7	1	1	1	1	1	1	1	1	1	1
	0.8	1	1	1	1	1	1	1	1	0	0
	0.9	1	0	0	0	0	0	0	0	0	0

	$d_1^{opt.}$ (Gy)	Δ_2: 0	0.1	0.2	0.3	0.4	0.5	0.6	0.7	0.8	0.9
Δ_1	0	0.00	0.00	0.00	0.00	0.00	0.00	0.00	0.00	0.00	0.00
	0.1	0.00	0.00	0.00	0.00	0.00	0.00	0.00	0.00	0.00	0.00
	0.2	0.00	0.00	0.00	0.00	0.00	0.00	0.00	0.00	0.00	0.00
	0.3	0.00	0.00	0.00	0.00	0.00	0.00	0.00	0.00	0.00	0.00
	0.4	0.00	0.00	0.00	0.00	0.00	0.00	0.00	2.87	2.69	2.41
	0.5	0.00	0.00	0.00	0.00	2.87	2.69	2.41	2.29	2.20	2.20
	0.6	0.00	0.00	2.69	2.41	2.29	2.20	2.11	2.03	2.03	1.96
	0.7	2.41	2.20	2.11	2.03	1.96	1.90	1.90	1.84	1.79	1.79
	0.8	1.90	1.84	1.79	1.75	1.75	1.70	1.66	1.66	1.52	1.52
	0.9	1.59	1.42	1.42	1.42	1.42	1.42	1.42	1.42	1.42	1.42

	$d_2^{opt.}$ (Gy)	Δ_2: 0	0.1	0.2	0.3	0.4	0.5	0.6	0.7	0.8	0.9
Δ_1	0	5.27	5.21	5.21	5.27	5.38	5.13	5.32	5.11	5.38	5.19
	0.1	5.18	5.15	5.18	5.27	5.01	5.18	5.43	5.21	5.01	5.36
	0.2	5.07	5.07	5.13	5.27	5.01	5.23	5.01	5.34	5.13	4.95
	0.3	5.31	5.39	5.08	5.27	5.01	5.31	5.08	5.52	5.31	5.11
	0.4	5.18	5.31	5.01	5.27	5.01	5.41	5.18	4.85	4.48	6.40
	0.5	5.01	5.21	4.91	5.27	4.89	4.48	6.36	5.88	5.44	5.24
	0.6	5.41	5.07	4.48	6.29	5.76	5.29	4.86	4.46	4.29	3.95
	0.7	6.19	5.37	4.86	4.40	4.00	3.63	3.46	3.15	2.87	2.75
	0.8	3.63	3.23	2.87	2.56	2.41	2.15	1.92	1.83	0.00	0.00
	0.9	1.38	0.00	0.00	0.00	0.00	0.00	0.00	0.00	0.00	0.00

	%	Δ_2: 0	0.1	0.2	0.3	0.4	0.5	0.6	0.7	0.8	0.9
Δ_1	0	0	9	16	22	28	33	37	40	44	47
	0.1	9	18	24	30	35	39	43	46	49	52
	0.2	19	27	33	38	42	46	49	52	55	57
	0.3	29	36	41	45	49	52	55	58	60	62
	0.4	39	45	49	53	56	59	62	63	64	65
	0.5	49	54	57	61	63	64	65	65	66	66
	0.6	59	63	64	65	66	66	67	67	67	68
	0.7	65	66	67	67	67	68	68	68	69	69
	0.8	68	68	69	69	69	69	69	70	70	70
	0.9	70	70	70	70	70	70	70	70	70	70

N_2^{optimal}, when $N_1^{\text{optimal}} = 0, d_1^{\text{optimal}} = 0$. Moreover, the corresponding optimal dose per session for the second modality can be simplified from (9.5) as

$$d_2^{\text{optimal}} = \frac{-\alpha_2 + \sqrt{(\alpha_2)^2 + \frac{4\beta_2(T\delta(\alpha_1+\beta_1\delta)(1-\Delta_1))}{(1+\Delta_2)N_2^{\text{optimal}}}}}{2s_2\beta_2}. \tag{9.20}$$

Since N_2^{optimal} is increasing in $N_2^\star$ (by an appropriate counterpart of Theorem 2.4), formulas (9.19) and (9.20) can help understand some of the qualitative trends in Table 9.1.

We enumerate several observations about the five panels within Table 9.1.

(a) Focus on the row corresponding to $\Delta_1 = 0$ in the second panel, which displays various N_2^{optimal} values. The values of N_2^{optimal} decrease as Δ_2 increases when we move right along this row. This can be explained by formula (9.19), because $N_2^\star$ is decreasing in Δ_2. Intuitively, the second modality becomes less desirable as Δ_2 increases. The same observation and explanation also hold for rows corresponding to $\Delta_1 \in \{0.1, 0.2, 0.3\}$, where $N_2^{\text{optimal}} > 0$, $N_1^{\text{optimal}} = 0$, and $d_1^{\text{optimal}} = 0$.

(b) When $\Delta_1 = 0.4$, the trend in (a) above continues across $\Delta_2 = 0 - 0.6$. As Δ_2 reaches 0.7, the second modality no longer dominates the first, and in fact, it is optimal to employ a combination of these two modalities (the first and second panels reveal that both $N_1^{\text{optimal}} > 0$ and $N_2^{\text{optimal}} > 0$ along with $d_1^{\text{optimal}} > 0, d_2^{\text{optimal}} > 0$ in the corresponding cells of the third and the fourth panels). In such combined therapy, the number of sessions administered with the first modality increases whereas the number of sessions with the second modality decreases, when Δ_2 increases.

(c) The switch from only using the second modality to using a combination of the two modalities as described in (b) above occurs at smaller values of Δ_2, as Δ_1 increases to 0.5 and 0.6. As Δ_1 becomes as high as 0.7, it is no longer optimal to use only the second modality. Specifically, it is optimal to use a combination of the two modalities even when Δ_2 is 0. For $\Delta_1 = 0.8$, it is in fact optimal to entirely abandon the second modality when Δ_2 reaches 0.8; this occurs at the much smaller value of $\Delta_2 = 0.1$ when $\Delta_1 = 0.9$.

(d) Consider the column corresponding to $\Delta_2 = 0$. The values of N_2^{optimal} decrease as Δ_1 increases when we move down along this column. When $N_1^{\text{optimal}} = 0$, this trend is also explained by formula (9.19), because $N_2^\star$ is decreasing in Δ_1. The same observation and explanation are also valid for all other columns when $N_1^{\text{optimal}} = 0$.

(e) The common theme in observations (b), (c), and (d) above is that, roughly speaking, the first modality becomes more desirable as Δ_1 increases. This is in contrast to the second modality, which becomes less desirable as Δ_2 increases. The apparent contradiction is perhaps rooted in the fact that dose-response parameters for the first modality appear on both sides of constraint (9.1) whereas those for the second modality don't. Thus, for example, when $N_1^{\text{optimal}} = 0$, the dose-response parameters that determine N_2^{optimal} are simply $\alpha_1^{\min} = \alpha_1(1-\Delta_1)$, $\beta_1^{\min}(1-\Delta_1)$, $\alpha_2^{\max} = \alpha_2(1+\Delta_2)$, and $\beta_2^{\max} = \beta_2(1+\Delta_2)$. Moreover, these parameters appear in formula (9.19) in a particularly simple manner that allows us to easily determine sensitivity with respect to Δ_2. On the contrary, even when $N_2^{\text{optimal}} = 0$, N_1^{optimal} is determined by dose-response parameters $\alpha_1^\star(k,\ell)$ and $\beta_1^\star(k,\ell)$ in subproblem (k,ℓ). Thus, the dependence of N_1^{optimal} on Δ_1 is theoretically not quite evident.

(f) Consider the third panel in Table 9.1, which displays various values of d_1^{optimal}. For each fixed value of Δ_1, wherever the values of $N_1^{\text{optimal}} > 0$ increase with Δ_2 as we move toward the right in the first panel, the corresponding values of $d_1^{\text{optimal}} > 0$ decrease in the third panel. Similarly, for each fixed value of Δ_2, wherever the values of $N_1^{\text{optimal}} > 0$ increase with Δ_1 as we move down in the first panel, the corresponding values of $d_1^{\text{optimal}} > 0$ decrease in the third panel. It is difficult to provide a full theoretical explanation for this trend, especially given that $N_2^{\text{optimal}} > 0$ and $d^{\text{optimal}} > 0$ in many of the relevant cells. Nevertheless, the trend is generally consistent with the basic idea that increasing the number of sessions forces the treatment planner to reduce the dose per session, to maintain the BED below the tolerable limit.

(g) Consider the fourth panel in Table 9.1, which displays various values of d_2^{optimal}. For any fixed Δ_1, the values of d_2^{optimal} exhibit an oscillating (up-and-down) trend as Δ_2 increases. Formula (9.20) suggests that this behavior is plausible (at least when $N_1^{\text{optimal}} = 0$) because while N_2^{optimal} decreases, $1+\Delta_2$ obviously increases, as Δ_2 increases. Similarly, for any fixed Δ_2, the values of d_2^{optimal} oscillate as Δ_1 increases. Formula (9.20) can explain this trend (at least when $N_1^{\text{optimal}} = 0$), again because while N_2^{optimal} decreases, $1-\Delta_1$ obviously decreases, when Δ_1 increases.

(h) Finally, we turn to the fifth panel in Table 9.1, which lists various percentage price of robustness values. The table shows that these values increase with both Δ_1 and Δ_2, as expected, similar to Chapters 4 and 7. This holds because the set of feasible (d_1, d_2) pairs for problem (2QR) with larger values of Δ_1, Δ_2 (and hence larger uncertainty intervals) is contained within the set of feasible (d_1, d_2) pairs with smaller values of Δ_1, Δ_2.

We are now ready to discuss the second set of experiments.

Table 9.2. *Results for Section 9.3.2 with different levels of uncertainty in the dose-response parameters of the two modalities. Decimals were suppressed in the bottom table labeled % to save space.*

	$N_1^{optimal}$	Δ_2 0	0.1	0.2	0.3	0.4	0.5	0.6	0.7	0.8	0.9
Δ_1	0	0	0	0	0	0	0	0	0	0	0
	0.1	0	0	0	0	0	0	0	0	0	0
	0.2	0	0	0	0	0	0	0	0	0	7
	0.3	0	0	0	0	0	0	0	8	9	10
	0.4	0	0	0	0	7	9	9	10	12	12
	0.5	0	0	7	9	10	12	12	13	14	14
	0.6	9	10	12	13	13	14	16	17	18	18
	0.7	13	14	16	17	18	18	19	19	20	20
	0.8	18	19	20	21	21	22	22	26	26	26
	0.9	24	29	29	29	29	29	29	29	29	29

	$N_2^{optimal}$	Δ_2 0	0.1	0.2	0.3	0.4	0.5	0.6	0.7	0.8	0.9
Δ_1	0	23	21	19	18	17	15	14	14	13	12
	0.1	21	19	17	16	15	14	13	12	12	11
	0.2	19	17	15	14	13	12	12	11	10	4
	0.3	16	15	14	12	12	11	10	4	3	3
	0.4	14	13	12	11	4	3	3	3	2	2
	0.5	12	11	4	3	3	2	2	2	2	2
	0.6	3	3	2	2	2	2	1	1	1	1
	0.7	2	2	1	1	1	1	1	1	1	1
	0.8	1	1	1	1	1	1	1	0	0	0
	0.9	1	0	0	0	0	0	0	0	0	0

	$d_1^{opt.}$(Gy)	Δ_2 0	0.1	0.2	0.3	0.4	0.5	0.6	0.7	0.8	0.9
Δ_1	0	0.00	0.00	0.00	0.00	0.00	0.00	0.00	0.00	0.00	0.00
	0.1	0.00	0.00	0.00	0.00	0.00	0.00	0.00	0.00	0.00	0.00
	0.2	0.00	0.00	0.00	0.00	0.00	0.00	0.00	0.00	0.00	2.87
	0.3	0.00	0.00	0.00	0.00	0.00	0.00	0.00	2.69	2.54	2.41
	0.4	0.00	0.00	0.00	0.00	2.87	2.54	2.54	2.41	2.20	2.20
	0.5	0.00	0.00	2.87	2.54	2.41	2.20	2.20	2.11	2.03	2.03
	0.6	2.54	2.41	2.20	2.11	2.11	2.03	1.90	1.84	1.79	1.79
	0.7	2.11	2.03	1.90	1.84	1.79	1.79	1.75	1.75	1.70	1.70
	0.8	1.79	1.75	1.70	1.66	1.66	1.62	1.62	1.52	1.52	1.52
	0.9	1.55	1.42	1.42	1.42	1.42	1.42	1.42	1.42	1.42	1.42

	$d_2^{opt.}$(Gy)	Δ_2 0	0.1	0.2	0.3	0.4	0.5	0.6	0.7	0.8	0.9
Δ_1	0	1.76	1.76	1.77	1.74	1.72	1.79	1.80	1.72	1.74	1.77
	0.1	1.74	1.75	1.78	1.75	1.74	1.74	1.75	1.78	1.71	1.75
	0.2	1.72	1.74	1.79	1.78	1.78	1.79	1.71	1.74	1.79	1.68
	0.3	1.77	1.73	1.71	1.80	1.71	1.73	1.77	1.58	1.80	1.66
	0.4	1.74	1.72	1.71	1.72	1.70	1.84	1.75	1.61	1.90	1.83
	0.5	1.71	1.70	1.67	1.78	1.62	1.90	1.82	1.66	1.52	1.46
	0.6	1.84	1.64	1.90	1.72	1.63	1.48	2.13	1.95	1.78	1.71
	0.7	1.69	1.50	2.13	1.92	1.73	1.64	1.49	1.42	1.28	1.23
	0.8	1.64	1.45	1.28	1.13	1.07	0.95	0.90	0.00	0.00	0.00
	0.9	0.64	0.00	0.00	0.00	0.00	0.00	0.00	0.00	0.00	0.00

	%	Δ_2 0	0.1	0.2	0.3	0.4	0.5	0.6	0.7	0.8	0.9
Δ_1	0	0	9	16	22	28	33	37	40	44	46
	0.1	9	18	24	30	35	39	43	46	49	52
	0.2	19	27	33	38	42	46	49	52	55	56
	0.3	29	36	41	45	49	52	55	57	57	58
	0.4	39	45	49	53	56	57	58	59	59	60
	0.5	49	54	56	57	58	59	60	60	61	61
	0.6	57	58	59	60	60	61	61	62	62	62
	0.7	60	61	61	62	62	62	63	63	63	63
	0.8	62	63	63	63	63	64	64	64	64	64
	0.9	64	64	64	64	64	64	64	64	64	64

9.3.2 Alternative Modality with Superior Biological Power

In this section, we let the sparing factor of the second modality be identical to the first and fix $s_2 = s_1 = 1$ throughout without any uncertainty. We let $\alpha_2^0 = f_0\alpha_1^0$; $\beta_2^0 = f_0\beta_1^0$; $\alpha_2 = f_2\alpha_1$; and $\beta_2 = f_2\beta_1$. Here, $f_0 > 0$ and $f_2 > 0$ are parameters that characterize the biological power of the second modality. Specifically, if $f_0 > 1$ and $f_0 > f_2$, the second modality is biologically superior to the first. We fix $f_0 = 2$ and $f_2 = 1$ throughout and model the uncertainty in the OAR dose-response parameters of both modalities via intervals exactly as in Section 9.3.1. The results are listed in Table 9.2, wherein the five panels display N_1^{optimal}, N_2^{optimal}, d_1^{optimal}, d_2^{optimal}, and the percentage price of robustness, from top to bottom, respectively. The two modalities are physically identical in all experiments here. In the nominal case where $\Delta_1 = \Delta_2 = 0$, the second modality is biologically superior to the first. Thus, it is optimal to administer only the second modality. Consequently, $N_2^{\text{optimal}} > 0, d_2^{\text{optimal}} > 0$, but $N_1^{\text{optimal}} = 0, d_1^{\text{optimal}} = 0$. As the values of Δ_1 and Δ_2 change across the rows and columns, all qualitative trends in the table are similar to those in Table 9.1. Those trends can be explained similarly as in Section 9.3.1.

Bibliographic Notes

The problem formulation and its reformulation as a finite set of subproblems in this chapter are based on the doctoral dissertation [91] and the associated conference paper [92]. An extension of the KKT approach described in Section 8.1 of Chapter 8 was developed there for solving each of the subproblems. Extensive numerical experiments were also conducted. The extension of the single-variable transformation method from Section 8.2, as presented in Section 9.2, is new.

Exercises

Exercise 9.1 Repeat the numerical experiments in this chapter for $T_{\text{double}} = 6$ days and also for $\rho_1 = 1/5\ \text{Gy}^{-1}$. Discuss any qualitative trends you observe after comparing your results with those reported in Tables 9.1 and 9.2.

Exercise 9.2 Repeat the numerical experiments from Table 9.1, for the case wherein there is uncertainty in the sparing factor $s_2 = 0.5$ of the second modality. Recall that this could, for example, model the uncertainty in the location of the Bragg peak for protons. Specifically, model this uncertainty using intervals defined by $s^{\min} = s_2(1 - \Delta_s)$ and $s^{\max} = s_2(1 + \Delta_s)$, with $\Delta_s \in \{0.1, 0.3, 0.5\}$.

Exercise 9.3 Develop a variation of Algorithm 8 that employs the KKT approach to solve subproblems $(2QR1_{k,\ell}(N_1, N_2))$ instead of the single-variable approach. Compare your results with those reported in Section 9.3.

Exercise 9.4 Extend Algorithm 8 to the case of multiple OAR. Do the same for its KKT approach-based variation that you designed in the above exercise. Perform numerical experiments and discuss any qualitative trends you observe.

10
Directions for Future Work

We conclude this monograph by outlining some directions for future research.

The inverse fractionation approach in Chapter 5 focused on problem (Q(N)) from Chapter 3. That is, the number of treatment sessions, N, was fixed. The goal then was to find values of dose-response parameters $\alpha_0 > 0, \beta_0 > 0$, and $\rho_m > 0$, for $m \in \mathcal{M}$, that would make given doses $d_1^\star, \ldots, d_N^\star$ optimal to (Q(N)). A natural extension of this inverse problem would instead focus on problem (Q) from Chapter 3. There, the goal would be to find values of dose-response parameters $\alpha_0 > 0, \beta_0 > 0, T_{\text{double}} > 0$, and $\rho_m > 0$, for $m \in \mathcal{M}$, that would make a given number of sessions $N^\star$ and corresponding given doses $d_1^\star, \ldots, d_{N^\star}^\star$ optimal to (Q). It is not evident whether and how the closed-form exact solution approach from Chapter 5 can be generalized to this more difficult problem. If such a generalization is not possible, it may be necessary to design an iterative algorithm for exact or approximate solution.

It would also be interesting to consider a spatiotemporally integrated extension of the inverse fractionation problem from Chapter 5. The simplified forward formulation presented in Chapter 6 would provide a starting point for this work. A closed-form solution of the resulting inverse formulation does not appear possible. Thus, the efforts will most likely concentrate on devising iterative algorithms for approximate solution.

The spatiotemporally integrated methodology in Chapter 6 solved a particular significant simplification of the original formulation (SP). It may be helpful to design algorithms to solve other simplifications, or better yet, the original formulation (SP) itself. Simplifications that do not assume that intensity profiles $u_t \in \Re_+^k$ are invariant across treatment sessions $t = 1: N$ would be of special interest. The literature on such formulations is evolving [1, 5, 56, 122], but opportunities remain for further algorithmic work.

The optimization models in this monograph were all static. That is, decisions were made at the beginning of the treatment course; no new information was obtained and thus decisions were not updated over the treatment course. This is largely consistent with current treatment protocols. The advent of new functional imaging technologies has engendered the possibility of acquiring biological information about the tumor's response to radiation over the treatment course. This creates an opportunity to update treatment plans based on such images. The idea of adapting radiation intensity profiles based on information acquired from functional images taken over the treatment course was envisioned, for example, in [78, 82, 96, 107, 108, 113, 117, 135]. An initial mathematical vision for this approach was laid in the doctoral dissertation [76], associated papers [77, 78], and a tutorial [59]. Subsequent work in this area included models and solution methods for adapting intensity profiles to (i) the uncertain evolution of oxygen concentration in the cancerous region (lack of oxygen reduces tumor radiosensitivity) [104]; (ii) learned values of uncertain tumor-response parameters [8]; and (iii) treatment course-lengths that are determined on the fly [7]. Any potential benefits and clinical viability of such approaches were debated in a point-counterpoint format in [79]. Additional clinical and mathematical research is needed to fully explore the potential of biologically adaptive spatiotemporally integrated planning.

Limited rigorous insight is available into the structure of optimal solutions to the optimal fractionation problem with two modalities. For example, what are the necessary and/or sufficient conditions for the (strict) optimality of a single-dosage solution to problem $(2Q1(N_1, N_2))$ in Chapter 8? That is, is it possible that setting either $d_1 > 0, d_2 = 0$ or $d_1 = 0, d_2 > 0$ achieves a larger objective value than any feasible solution where both d_1 and d_2 are positive? What are the necessary and/or sufficient conditions under which this occurs? Similarly for problem $(2Q(N_1, N_2))$. More generally, what are the necessary and/or sufficient conditions for the strict optimality of $N_1^{\text{optimal}} > 0$, $N_2^{\text{optimal}} > 0$, $d_1^{\text{optimal}} > 0$, and $d_2^{\text{optimal}} > 0$ in problem (2Q)?

The formulation of the optimal fractionation problem with two modalities in Chapter 8 assumed that an identical dose is administered in each session with each modality. A more general approach would allow for doses to vary across sessions, for each modality. This would yield a formulation with nonnegative dose variables $d_1^1, \ldots, d_{N_1}^1$ for the first modality and $d_1^2, \ldots, d_{N_1}^2$ for the second modality. A direct extension of the solution methods from Chapter 3 may not be possible. New research into tailored iterative solution algorithms may therefore be needed.

An inverse fractionation problem with two modalities has not yet been studied in the literature. An extension of the closed-form exact solution method from Chapter 5 thus remains unexplored. If closed-form solution is not possible, a tailored iterative solution algorithm for exact or approximate solution may be needed.

A spatiotemporally integrated counterpart of the formulations in Chapter 6 in the context of two modalities is also not available. An efficient iterative solution algorithm for such formulations is also missing. Similarly for an extension of the spatiotemporally integrated robust optimization approach from Chapter 7 to the case of two modalities. Finally, an inverse optimization framework for spatiotemporally integrated fractionation with two modalities remains entirely unexplored. This provides opportunities for future algorithmic work.

More complicated variants of the LQ dose-response model have been studied in the literature [71, 74, 84, 109]. Other variants may also appear in the future. In fact, entirely different dose-response frameworks, perhaps unrelated to the LQ philosophy, may become mainstream in the more distant future. Corresponding formulations of the fractionation problem may call for new solution methodologies. We hope that the insights presented in this monograph will prove helpful in navigating that new frontier of research.

Bibliographic Notes

A brief summary of these future research directions was outlined in a tutorial [61].

Appendix

Background on Optimization

This appendix provides a brief informal introduction to some basic concepts in optimization. For a more rigorous and extensive treatment, we refer the reader to [18] for linear optimization and [17, 23] for nonlinear optimization. The reader is assumed to be familiar with linear algebra [115] and multivariate calculus [44].

A.1 Linear Programming

A linear program (LP) is an optimization problem that can be expressed as

$$\max\ c_1x_1 + \cdots + c_nx_n \tag{A.1}$$

$$a_{i1}x_1 + \cdots + a_{in}x_n \leq b_i,\ i = 1, \ldots, m \tag{A.2}$$

$$x_1 \geq 0, \ldots, x_n \geq 0. \tag{A.3}$$

Here, $x_1, \ldots, x_n$ are real-valued decision variables. The objective in (A.1) is to maximize a linear function of these decision variables. The given real numbers $c_1, \ldots, c_n$ are called coefficients of this linear objective function. The decision variables are restricted to be nonnegative in (A.3). The problem includes m other linear constraints (A.2), indexed $i = 1, \ldots, m$. The right-hand side b_i of the ith constraint is a given real number. Coefficients $a_{i1}, \ldots, a_{in}$ of the linear function on the left-hand side in the ith constraint are also given real numbers. More generally, an LP may call for minimizing a linear function, may include linear constraints of the forms $\geq$ or $=$, and may include variables that are unrestricted in sign or are restricted to be nonpositive. In fact, all such alternative LPs can be converted into an equivalent problem of the form (A.1)–(A.3).

Three distinct possibilities can arise when one attempts to solve an LP of the form (A.1)–(A.3).

First, there may not be any values of $x_1, \ldots, x_n$ that satisfy the constraints in an LP. This occurs, for instance, for the LP $\max x_1$ subject to constraints $x_1 \leq -1$ and $x_1 \geq 0$. If decision variable values $x_1, \ldots, x_n$ satisfy all constraints in the LP, then $x_1, \ldots, x_n$ is called a feasible solution. In that case, the LP is said to be feasible. The set of all feasible solutions to an LP is called the feasible region. Such a feasible region is a subset of the Euclidean space $\mathbb{R}^n$. It is called a polyhedron.

Second, the objective function value of an LP may be unbounded. That is, for any real number R, there exist decision variable values $x_1, \ldots, x_n$ that satisfy all constraints in the LP and $c_1x_1 + \cdots + c_nx_n > R$. This happens, for instance, in the LP max x_1 subject to constraints $x_1 \geq 1$ and $x_1 \geq 0$.

The third possibility is that there exist $x_1^\star, \ldots, x_n^\star$ that satisfy all constraints in the LP and $c_1x_1^\star + \cdots + c_nx_n^\star \geq c_1u_1 + \cdots + c_nu_n$ for any decision variable values $u_1, \ldots, u_n$ that satisfy all constraints in the LP. In that case, $x_1^\star, \ldots, x_n^\star$ is called an optimal solution to (or a maximizer of) the LP. This happens for the LP max x_1 subject to constraints $x_1 \leq 1$ and $x_1 \geq 0$, where the solution $x_1^\star = 1$ is optimal. In general, an LP can have multiple optimal solutions. If an LP has an optimal solution, then at least one optimal solution must occur at a corner point of the feasible region.

Optimal solutions to LPs with two decision variables can be identified by superimposing objective function contour lines, that is, lines along which the objective function attains constant values, on a plot of the feasible region. Consider, for example, the LP

$$\max \; x_1 + x_2 \tag{A.4}$$

$$2x_1 + x_2 \leq 12 \tag{A.5}$$

$$x_1 + 2x_2 \leq 9 \tag{A.6}$$

$$x_1 \geq 0, \; x_2 \geq 0. \tag{A.7}$$

This LP has the single optimal solution $x_1^\star = 5$ and $x_2^\star = 2$ as identified in Figure A.1.

More generally, an LP with more than two variables can be solved using specialized algorithms such as the Simplex method or the interior point method.

Each LP is associated with another LP called the dual problem. To illustrate, we consider the following extensive form LP

$$\begin{aligned}
\max \; & c_1x_1 + \cdots + c_nx_n \\
& \sum_{j=1}^{n} a_{ij}x_j = b_i, \; i \in I_1 \\
& \sum_{j=1}^{n} a_{ij}x_j \geq b_i, \; i \in I_2 \\
& \sum_{j=1}^{n} a_{ij}x_j \leq b_i, \; i \in I_3 \\
& x_j \text{ free}, \; j \in J_1 \\
& x_j \geq 0, \; j \in J_2 \\
& x_j \leq 0, \; j \in J_3.
\end{aligned}$$

Here, J_1, J_2, J_3 are exhaustive and mutually exclusive subsets of the index set $\{1, \ldots, n\}$ of decision variables $x_1, \ldots, x_n$; and I_1, I_2, I_3 are exhaustive and mutually exclusive subsets of the index set $\{1, \ldots, m\}$ of constraints. We consider this LP here because it explicitly includes all possible types of constraints and variable signs, unlike the original LP (P). We call this the primal problem. The dual of this LP is given by

$$\min \; b_1y_1 + \cdots + b_my_m$$

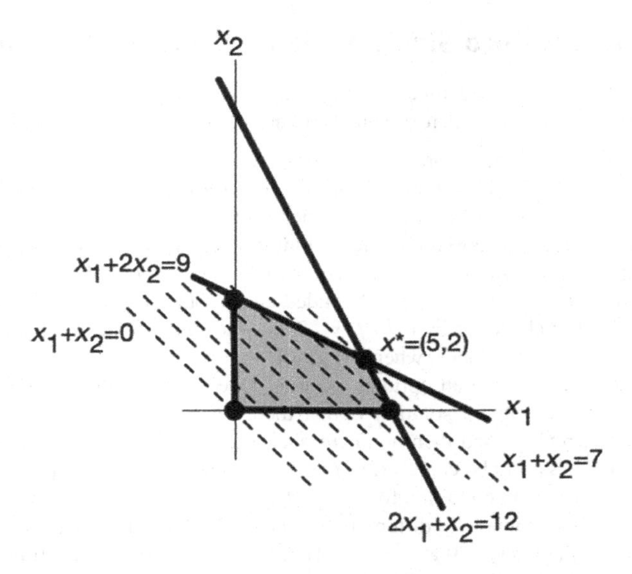

Figure A.1 Constraint boundaries $x_1 = 0$, $x_2 = 0$, $2x_1 + x_2 = 12$, and $x_1 + 2x_2 = 9$ are shown as thick black lines. Feasible region of LP (A.4)–(A.7) is shown in shaded gray. All corner points of this feasible region are shown as black dots. Contour lines of the objective function, where $x_1 + x_2$ equals distinct constant values such as 0 and 7, are shown as dashed lines. Since the goal is to maximize $x_1 + x_2$, we look for a contour line that is as far away out from the origin $(0, 0)$, without leaving the feasible region. This helps us identify the optimal corner point solution $x_1^\star = 5$, $x_2^\star = 2$.

$$\sum_{i \in I_1} a_{ij} y_i = c_i, \; j \in J_1$$

$$\sum_{i \in I_2} a_{ij} y_i \geq c_i, \; j \in J_2$$

$$\sum_{i \in I_3} a_{ij} y_i \leq c_i, \; j \in J_3$$

$$y_i \text{ free}, \; i \in I_1$$

$$y_i \leq 0, \; i \in I_2$$

$$y_i \geq 0, \; i \in I_3.$$

These two problems exhibit several useful relationships. For instance, if $x_1, \ldots, x_n$ is feasible to the primal problem and $y_1, \ldots, y_m$ is feasible to the dual problem, then $\sum_{i=1}^m b_i y_i \geq \sum_{j=1}^n c_j x_j$. This is called weak duality. Moreover, one problem has an optimal solution if, and only if, the other problem does; and in that case, the optimal objective function values in the two problems are equal. This is called strong duality.

A.2 Unconstrained Single-Variable Nonlinear Optimization

Suppose f is a real-valued function of a single decision variable x. Consider the problem $\max_{x \in \mathbb{R}} f(x)$. Here, x is real valued and has no (other) constraints. We call this a single-variable unconstrained optimization problem.

We say that $x^\star \in \mathbb{R}$ is an optimal solution to (or a maximizer of) this problem if $f(x^\star) \geq f(y)$, for any $y \in \mathbb{R}$. Two distinct possibilities can occur.

The problem may not have an optimal solution. There are in turn two ways in which this can happen. First, the function f may be unbounded above. For example, when $f(x) = x$. Second, the function f is bounded above, but for any $\hat{x} \in \mathbb{R}$, there exists another $x^\star \in \mathbb{R}$ such that $f(x^\star) > f(\hat{x})$. Thus, the function never attains a largest value. This occurs, for instance, when $f(x) = 1 - \exp(-x)$.

The problem may have an optimal solution. One evident sufficient condition for the existence of an optimal solution is that there exists an $x^\star \in \mathbb{R}$ such that f is nondecreasing up to $x^\star$ and then nonincreasing after $x^\star$. That is, f is quasiconcave. The function f is said to be coercive, if $f(x)$ approaches $-\infty$ as $|x|$ approaches $+\infty$. Coercivity is a sufficient condition for the existence of an optimal solution.

Suppose f is continuously differentiable on $\mathbb{R}$. If $x^\star$ is an optimal solution, then we must have $df(x)/dx = 0$ at $x = x^\star$. That is, the slope of the function must be 0 at an optimal solution. This is called a first order necessary condition for optimality. Thus, when an optimal solution is known to exist, one can in principle identify it by first finding all x where the slope is 0 and then comparing the objective function values across all such x.

The function f is said to be concave on $\mathbb{R}$ if it satisfies the following property. Consider any $x, y \in \mathbb{R}$. Then, for any real number $0 \leq \lambda \leq 1$, we have $\lambda f(x) + (1-\lambda) f(y) \leq f(\lambda x + (1-\lambda) y)$. That is, the function value at any point between x and y is at least as large as the corresponding convex combination of function values at x and y. Geometrically, the chord connecting $(x, f(x))$ and $(y, f(y))$ lies below the graph of the function. See Figure A.2. A function f is said to be convex on $\mathbb{R}$ if $-f$ is concave. Suppose f is concave and continuously differentiable on $\mathbb{R}$. Suppose $df(x)/dx = 0$ at $x = x^\star$. Then, $x^\star$ is optimal. This is called a first order sufficient condition for optimality of such a concave function. Suppose f is twice continuously differentiable on $\mathbb{R}$. Then, f is concave on $\mathbb{R}$ if, and only if, $d^2 f(x)/dx^2 \leq 0$, for all $x \in \mathbb{R}$. That is, the first derivative itself is a decreasing function. This is consistent with the intuition that a concave function curves downwards. A strict inequality < 0 for the second derivative here characterizes a strictly concave function.

Some of these ideas can be extended to unconstrained problems where the objective f is a function of n decision variables $x_1, \ldots, x_n$. We omit that discussion here.

A.3 Constrained Nonlinear Optimization with Multiple Variables

Consider the optimization problem

$$(P) \ \max \ f(x_1, \ldots, x_n)$$

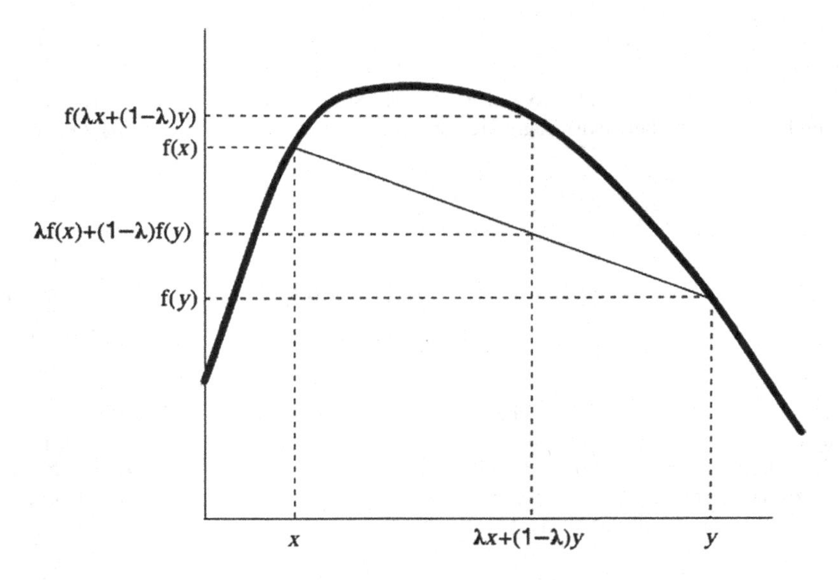

Figure A.2 A concave function of a single variable is shown as a thick black line. The chord shown as a thin black line lies below the function.

$$g_i(x_1, \ldots, x_n) \leq 0, \; i = 1, \ldots, m$$
$$h_j(x_1, \ldots, x_n) = 0, \; j = 1, \ldots, \ell.$$

Assume that the functions $f\colon \mathbb{R}^n \to \mathbb{R}$; $g_i\colon \mathbb{R}^n \to \mathbb{R}$, for $i = 1, \ldots, m$; and $h_j\colon \mathbb{R}^n \to \mathbb{R}$, for $j = 1, \ldots, \ell$, are continuously differentiable. As such, this problem includes n decision variables $x_1, \ldots, x_n$. The problem is nonlinear when at least one of the functions is nonlinear. Conversely, the problem reduces to an LP if all functions are linear. The problem is called a quadratically constrained quadratic program (QCQP) when all functions are quadratic. When f is quadratic but g_i and h_j are linear, the problem is called a quadratic program (QP). The problem is convex if function f is concave; functions g_i, for $i = 1, \ldots, m$, are convex; and functions h_j, for $j = 1, \ldots, \ell$, are linear.

The problem reduces to an unconstrained one, when $m = \ell = 0$. We will use ∇f to denote the gradient, that is, the n-dimensional vector of partial derivatives of f with respect to $x_1, \ldots, x_n$. Similarly for ∇g_i and ∇h_j.

We say that $\vec{x} = (x_1, \ldots, x_n) \in \mathbb{R}^n$ is feasible to problem (P), if $g_i(x_1, \ldots, x_n) \leq 0$, for $i = 1, \ldots, m$; and $h_j(x_1, \ldots, x_n) = 0$, for $j = 1, \ldots, \ell$. The set of all feasible solutions is called the feasible region. We denote it by $S \subseteq \mathbb{R}^n$.

Suppose $\vec{x}^\star = (x_1^\star, \ldots, x_n^\star) \in S$ is a feasible solution to (P). Suppose there exists a real number $r > 0$ such that $f(\vec{x}^\star) \geq f(\vec{x})$ for any feasible solution $\vec{x} \in S \cap B_r(\vec{x}^\star)$. Then, $\vec{x}^\star$ is called a local maximizer of (P). Here, $B_r(\vec{x}^\star) = \{\vec{x} \in \mathbb{R} \,|\, ||\vec{x}^\star - \vec{x}||_2 \leq r\}$ denotes the Euclidean ball of radius r around $\vec{x}^\star$. Intuitively, the objective function value of $\vec{x}^\star$ is at least as large as that of any feasible solution in some ball around $\vec{x}^\star$.

We say that $\vec{x}^\star \in S$ is an optimal solution to (or maximizer of) (P) if $f(\vec{x}^\star) \geq f(\vec{x})$, for all $\vec{x} \in S$.

Suppose $\vec{x}^\star$ is a local maximizer of (P). Suppose $\vec{x}$ satisfies some regularity conditions. Then, there exist Lagrange multipliers $\lambda_1, \ldots, \lambda_m$ and $\mu_1, \ldots, \mu_\ell$ such that

$$\sum_{i=1}^{m} \lambda_i \nabla g_i(\vec{x}^\star) + \sum_{j=1}^{\ell} \mu_j \nabla h_j(\vec{x}^\star) = \nabla f(\vec{x}^\star) \tag{A.8}$$

$$g_i(\vec{x}^\star) \leq 0, \; i = 1, \ldots, m \tag{A.9}$$

$$h_j(\vec{x}^\star) = 0, \; j = 1, \ldots, \ell \tag{A.10}$$

$$\lambda_i g_i(\vec{x}^\star) = 0, \; i = 1, \ldots, m \tag{A.11}$$

$$\lambda_i \geq 0, \; i = 1, \ldots, m. \tag{A.12}$$

These are the Karush–Kuhn–Tucker (KKT) first order necessary conditions for (local) optimality. In this book we have ignored the regularity conditions for the most part, thus implicitly excluding the case where all local maxima are irregular. Thus, when a maximizer is known to exist, one can in principle identify it by first finding all $\vec{x}$ that satisfy these KKT conditions and then comparing objective function values across all such $\vec{x}$.

References

[1] Adibi, A, and Salari, E. 2018. Spatiotemporal radiotherapy planning using a global optimization approach. *Physics in Medicine and Biology*, **63**, 035040.

[2] Adibi, A, and Salari, E. 2021. Scalable optimization methods for incorporating spatiotemporal fractionation into IMRT planning. Forthcoming in *INFORMS Journal on Computing*.

[3] Ahamad, A, Rosenthal, D I, and Ang, K K. 2005. *Squamous cell head and neck cancer: Recent clinical progress and prospects for the future*. Current Clinical Oncology. Totowa, NJ: Humana Press.

[4] Ajdari, A. 2017. Robust, non-stationary, and adaptive fractionation in radiotherapy. PhD thesis, University of Washington, Seattle.

[5] Ajdari, A, and Ghate, A. 2016. A model predictive control approach for discovering nonstationary fluence-maps in cancer radiotherapy fractionation. Pages 2065–2075 of: *Proceedings of the Winter Simulation Conference*. Arlington, VA: IEEE Press.

[6] Ajdari, A, and Ghate, A. 2016. Robust spatiotemporally integrated fractionation in radiotherapy. *Operations Research Letters*, **44**(4), 544–549.

[7] Ajdari, A, Ghate, A, and Kim, M. 2018. Adaptive treatment-length optimization in spatiobiologically integrated radiotherapy. *Physics in Medicine and Biology*, **63**(7), 075009.

[8] Ajdari, A, Saberian, F, and Ghate, A. 2020. A theoretical framework for learning tumor dose-response uncertainty in individualized spatiobiologically integrated radiotherapy. *INFORMS Journal on Computing*, **32**(4), 930–951.

[9] Aleman, D M. 2018. Fluence map optimization in intensity-modulated radiation therapy treatment planning. Pages 285–306 of: Kong, N, and Zhang, S (eds), *Decision analytics and optimization in disease prevention and treatment*. Hoboken, NJ: John Wiley & Sons.

[10] Arcangeli, G, Saracino, B, Gomellini, S, Petrongari, M G, Arcangeli, S, Sentinelli, S, Marzi, S, Landoni, V, Fowler, J, and Strigari, L. 2010. A prospective phase III randomized trial of hypofractionation versus conventional fractionation in patients with high-risk prostate cancer. *International Journal of Radiation Oncology*Biology*Physics*, **78**(1), 11–18.

[11] Armpilia, C I, Dale, R G, and Jones, B. 2004. Determination of the optimum dose per fraction in fractionated radiotherapy when there is delayed onset of tumour repopulation during treatment. *The British Journal of Radiology*, **77**(921), 765–767.

[12] Auckly, D. 2007. Solving the quartic with a pencil. *The American Mathematical Monthly*, **114**(1), 29–39.

[13] Badri, H, Watanabe, Y, and Leder, K. 2016. Optimal radiotherapy dose schedules under parametric uncertainty. *Physics in Medicine and Biology*, **61**(1), 338–364.

[14] Ben-Tal, A, El Ghaoui, L, and Nemirovski, A. 2009. *Robust optimization*. Applied Mathematics. Princeton, NJ: Princeton University Press.

[15] Benjamin, L C, Tree, A C, and Dearnaley, D P. 2017. The role of hypofractionated radiotherapy in prostate cancer. *Current Oncology Reports*, **19**(4), 1–9.

[16] Bernier, J, and Horiot, J-C. 2012. Altered-fractionated radiotherapy in locally advanced head and neck cancer. *Current Opinion in Oncology*, **24**(3), 223–228.

[17] Bertsekas, D P. 2008. *Nonlinear programming*. Belmont, MA: Athena Scientific.

[18] Bertsimas, D, and Tsitsiklis, J. 1997. *Introduction to linear optimization*. Belmont, MA: Athena Scientific.

[19] Bertsimas, D, Brown, D B, and Caramanis, C. 2011. Theory and applications of robust optimization. *SIAM Review*, **53**(3), 464–501.

[20] Bertuzzi, A, Bruni, C, Papa, F, and Sinisgalli, C. 2013. Optimal solution for a cancer radiotherapy problem. *Journal of Mathematical Biology*, **66**(1–2), 311–349.

[21] Bortfeld, T, Chan, T C Y, Trofimov, A, and Tsitsiklis, J N. 2008. Robust management of motion uncertainty in intensity modulated radiation therapy. *Operations Research*, **56**(6), 1461–1473.

[22] Bortfeld, T, Ramakrishnan, J, Tsitsiklis, J N, and Unkelbach, J. 2015. Optimization of radiotherapy fractionation schedules in the presence of tumor repopulation. *INFORMS Journal on Computing*, **27**(4), 788–803.

[23] Boyd, S, and Vandenberghe, L. 2004. *Convex optimization*. Cambridge: Cambridge University Press.

[24] Brenner, D J. 2008. Point: The linear-quadratic model is an appropriate methodology for determining iso-effective doses at large doses per fraction. *Seminars in Radiation Oncology*, **18**(4), 234–239.

[25] Brenner, D J, and Hall, E J. 2018. Hypofractionation in prostate cancer radiotherapy. *Translational Cancer Research*, **7**(Supplement 6), S632–S639.

[26] Brenner, D J, and Herbert, D E. 1997. The use of the linear-quadratic model in clinical radiation oncology can be defended on the basis of empirical evidence and theoretical argument. *Medical Physics*, **24**(8), 1245–1248.

[27] Brown, A, and Suit, H. 2004. The centenary of the discovery of the Bragg peak. *Radiotherapy and Oncology*, **73**(3), 265–268.

[28] Burman, C, Chui, C-S, Kutcher, G, Leibel, S, Zelefsky, M, LoSasso, T, Spirou, S, Wu, Q, Yang, J, Stein, J, Mohan, R, Fuks, Z, and Ling, C C. 1997. Planning, delivery, and quality assurance of intensity-modulated radiotherapy using dynamic multileaf collimator: A strategy for large-scale implementation for the treatment of carcinoma of the prostate. *International Journal of Radiation Oncology*Biology*Physics*, **39**(4), 863–873.

[29] Catton, C, and Lukka, H. 2019. The evolution of fractionated prostate cancer radiotherapy. *The Lancet*, **394**(10196), 361–362.

[30] Chan, T C Y, and Mišić, V V. 2013. Adaptive and robust radiation therapy optimization for lung cancer. *European Journal of Operational Research*, **231**(3), 745–756.

[31] Chan, T C Y, Bortfeld, T, and Tsitsiklis, J N. 2006. A robust approach to IMRT optimization. *Physics in Medicine and Biology*, **51**(10), 2567–2583.

[32] Chan, T C Y, Mahmood, R, and Zhu, I Y. 2021 (September). *Inverse optimization: Theory and applications*. https://arxiv.org/abs/2109.03920.

[33] Cho, B. 2018. Intensity-modulated radiation therapy: A review with a physics perspective. *Radiation Oncology Journal*, **36**(1), 1–10.

[34] Craft, D, Bangert, M, Long, T, Papp, D, and Unkelbach, J. 2014. Shared data for intensity modulated radiation therapy (IMRT) optimization research: The CORT dataset. *Gigascience*, **3**(1), 37.

[35] Dale, R G. 1985. The application of the linear-quadratic dose-effect equation to fractionated and protracted radiotherapy. *British Journal of Radiology*, **58**(690), 515–528.

[36] Dasu, A, and Toma-Dasu, I. 2012. Prostate alpha/beta revisited: An analysis of clinical results from 14 168 patients. *Acta Oncologica*, **51**(8), 963–974.

[37] Deasy, J, Blanco, A, and Clark, V. 2003. CERR: A computational environment for radiotherapy research. *Medical Physics*, **30**(5), 979–985.

[38] DeLaney, T F. 2011. Proton therapy in the clinic. *Frontiers of Radiation Therapy and Oncology*, **43**, 465–485.

[39] Duchesne, G M, and Peters, L J. 1999. What is the α/β ratio for prostate cancer? Rationale for hypofractionated high-dose-rate brachytherapy. *International Journal of Radiation Oncology*Biology*Physics*, **44**(4), 747–748.

[40] Ebner, D K, and Kamada, T. 2016. The emerging role of carbon-ion radiotherapy. *Frontiers in Oncology*, **6**(140), 1–6.

[41] Ehrgott, M, Guler, C, Hamacher, H W, and Shao, L. 2008. Mathematical optimization in intensity modulated radiation therapy. *4OR*, **6**(3), 199–262.

[42] Eisbruch, A. 2002. Intensity-modulated radiotherapy of head-and-neck cancer: Encouraging early results. *International Journal of Radiation Oncology*Biology*Physics*, **53**(1), 1–3.

[43] Ferris, M C, and Voelker, M M. 2004. Fractionation in radiation treatment planning. *Mathematical Programming*, **101**, 387–413.

[44] Fitzpatrick, P M. 2009. *Advanced calculus*. Providence, RI: American Mathematical Society.

[45] Fowler, J F. 1984. Non-standard fractionation in radiotherapy. *International Journal of Radiation Oncology*Biology*Physics*, **10**(5), 755–759.

[46] Fowler, J F. 1989. The linear-quadratic formula and progress in fractionated radiotherapy. *The British Journal of Radiology*, **62**(740), 679–694.

[47] Fowler, J F. 1990. How worthwhile are short schedules in radiotherapy?: A series of exploratory calculations. *Radiotherapy and Oncology*, **18**(2), 165–181.

[48] Fowler, J F. 1992. Brief summary of radiobiological principles in fractionated radiotherapy. *Seminars in Radiation Oncology*, **2**(1), 16–21.

[49] Fowler, J F. 2001. Biological factors influencing optimum fractionation in radiation therapy. *Acta Oncologica*, **40**(6), 712–717.

[50] Fowler, J F. 2007. Is there an optimal overall time for head and neck radiotherapy? A review with new modeling. *Clinical Oncology*, **19**(1), 8–27.

[51] Fowler, J F. 2008. Optimum overall times II: Extended modelling for head and neck radiotherapy. *Clinical Oncology*, **20**(2), 113–126.

[52] Fowler, J F. 2010. 21 years of biologically effective dose. *British Journal of Radiology*, **83**(991), 554–568.

[53] Fowler, J F, Chappell, R, and Ritter, M. 2001. Is alpha/beta for prostate tumors really low? *International Journal of Radiation Oncology*Biology*Physics*, **50**(4), 1021–1031.

[54] Fu, K K, Pajak, T F, Trotti, A, Jones, C U, Spencer, S A, Phillips, T L, Garden, A S, Ridge, J A, Cooper, J S, and Ang, K K. 2000. A Radiation Therapy Oncology Group (RTOG) phase III randomized study to compare hyperfractionation and two variants of accelerated fractionation to standard fractionation radiotherapy for head and neck squamous cell carcinomas: First report of RTOG 9003. *International Journal of Radiation Oncology*Biology*Physics*, **48**(1), 7–16.

[55] Gabrel, V, Murat, C, and Thiele, A. 2014. Recent advances in robust optimization: An overview. *European Journal of Operational Research*, **235**(3), 471–483.

[56] Gaddy, M R, Yildiz, S, Unkelbach, J, and Papp, D. 2018. Optimization of spatiotemporally fractionated radiotherapy treatments with bounds on the achievable benefit. *Physics in Medicine and Biology*, **63**, 015036.

[57] Garcia, L M, Leblanc, J, Wilkins, D, and Raaphorst, G P. 2006. Fitting the linear–quadratic model to detailed data sets for different dose ranges. *Physics in Medicine and Biology*, **51**(11), 2813–2823.

[58] Garden, A S. 2001. Altered fractionation for head and neck cancer. *Oncology*, **15**(10), 1326–1332.

[59] Ghate, A. 2011. Dynamic optimization in radiotherapy. Pages 60–74 of: Gray, P (ed), *INFORMS TutORials in Operations Research*. Catonsville, MD: Institute for Operations Research and the Management Sciences. https://pubsonline.informs.org/doi/pdf/10.1287/educ.1110.0088. Catonsville.

[60] Ghate, A. 2020. Imputing radiobiological parameters of the linear-quadratic dose-response model from a radiotherapy fractionation plan. *Physics in Medicine and Biology*, **65**(22), 225009.

[61] Ghate, A. 2021. Response-guided dosing in cancer radiotherapy. Pages 1–37 of: Gray, P (ed), *INFORMS TutORials in Operations Research*. Catonsville, MD: Institute for Operations Research and the Management Sciences.

[62] Gorissen, B L. 2019. Guaranteed ε-optimal solutions with the linear optimizer ART3+O. *Physics in Medicine and Biology*, **64**(7), 075017.

[63] Greco, C, Pimentel, N, Pares, O, Louro, V, and Fuks, Z Y. 2018. Single-dose radiotherapy (SDRT) in the management of intermediate risk prostate cancer: Early results from a phase II randomized trial. *Journal of Clinical Oncology*, **36**(6), 128.

[64] Grutters, J P, Kessels, A G H, Pijls-Johannesma, M, de Ruysscher, D, Joore, M A, and Lambin, P. 2010. Comparison of the effectiveness of radiotherapy

with photons, protons and carbon-ions for non-small cell lung cancer: A meta-analysis. *Radiotherapy and Oncology*, **95**(1), 32–40.

[65] Hall, E J, and Giaccia, A J. 2005. *Radiobiology for the radiologist*. Philadelphia, PA: Lippincott Williams & Wilkins.

[66] Halperin, E C. 2006. Particle therapy and treatment of cancer. *Lancet Oncology*, **7**(8), 676–685.

[67] Halperin, E C, Brady, L W, and Perez, C A. 2013. *Perez and Brady's principles and practice of radiation oncology*. Philadelphia, PA: Lippincott Williams & Wilkins (Wolters Kluwer Health).

[68] Ho, K F, Fowler, J F, Sykes, A J, Yap, B K, Lee, L W, and Slevin, N J. 2009. IMRT dose fractionation for head and neck cancer: Variation in current approaches will make standardisation difficult. *Acta Oncologica*, **48**(3), 431–439.

[69] Horiot, J C, Le Fur, R, N'Guyen, T, Chenal, C, Schraub, S, Alfonsi, S, Gardani, G, van den Bogaert, W, Danczak, S, and Bolla, M. 1992. Hyperfractionation versus conventional fractionation in oropharyngeal carcinoma: Final analysis of a randomized trial of the EORTC cooperative group of radiotherapy. *Radiotherapy Oncology*, **25**(4), 231–241.

[70] Horiot, J C, Bontemps, P, van den Bogaert, W, Fur, R L, van den Weijngaert, D, Bolla, M, Bernier, J, Lusinchi, A, Stuschke, M, Lopez-Torrecilla, J, Begg, A C, Pierart, M, and Collette, L. 1997. Accelerated fractionation (AF) compared to conventional fractionation (CF) improves loco-regional control in the radiotherapy of advanced head and neck cancers: Results of the EORTC 22851 Randomized Trial. *Radiotherapy and Oncology: Journal of the European Society for Therapeutic Radiology and Oncology*, **44**(2), 111–121.

[71] Jeong, J, Oh, J H, Sonke, J-J, Belderbos, J, Bradley, J D, Fontanella, A N, Rao, S S, and Deasy, J O. 2017. Modeling the cellular response of lung cancer to radiation therapy for a broad range of fractionation schedules. *Clinical Cancer Research*, **23**(18), 5469–5479.

[72] Jones, B, Tan, L T, and Dale, R G. 1995. Derivation of the optimum dose per fraction from the linear quadratic model. *The British Journal of Radiology*, **68**(812), 894–902.

[73] Jones, L, Hoban, P, and Metcalfe, P. 2001. The use of the linear quadratic model in radiotherapy: A review. *Australasian Physical & Engineering Sciences in Medicine*, **24**(3), 132–146.

[74] Kehwar, T S, Chopra, K L, and Rai, D V. 2017. A unified dose response relationship to predict high dose fractionation response in the lung cancer stereotactic body radiation therapy. *Journal of Medical Physics*, **42**(4), 222–233.

[75] Keller, H, Meier, G, Hope, A, and Davison, M. 2012. Fractionation schedule optimization for lung cancer treatments using radiobiological and dose distribution characteristics. *Medical Physics*, **39**(6), 3811–3811.

[76] Kim, M. 2010. A mathematical framework for spatiotemporal optimality in radiation therapy. PhD thesis, University of Washington, Seattle.

[77] Kim, M, Ghate, A, and Phillips, M H. 2009. A Markov decision process approach to temporal modulation of dose fractions in radiation therapy planning. *Physics in Medicine and Biology*, **54**(14), 4455–4476.

[78] Kim, M, Ghate, A, and Phillips, M H. 2012. A stochastic control formalism for dynamic biologically conformal radiation therapy. *European Journal of Operational Research*, **219**(3), 541–556.

[79] Kim, M, Craft, D, and Orton, C G. 2016. Within the next five years, most radiotherapy treatment schedules will be designed using spatiotemporal optimization. *Medical Physics*, **43**(5), 2009–2012.

[80] Kirkpatrick, J P, Meyer, J J, and Marks, L B. 2008. The linear-quadratic model is inappropriate to model high dose per fraction effects in radiosurgery. *Seminars in Radiation Oncology*, **18**(4), 240–243.

[81] Laboratory, Brookhaven National. 2022. *NASA Space Research Laboratory User Guide*. www.bnl.gov/nsrl/userguide/bragg-curves-and-peaks.php.

[82] Lawrence, Y R, Werner-Wasik, M, and Dicker, A P. 2008. Biologically conformal treatment: Biomarkers and functional imaging in radiation oncology. *Future Oncology*, **4**(5), 689–704.

[83] Lea, D E, and Catcheside, D G. 1942. The mechanism of the induction by radiation of chromosome aberrations in Tradescantia. *Journal of Genetics*, **44**(2–3), 216–245.

[84] Li, S. 2021. Theoretical derivation and clinical dose-response quantification of a unified multi-activation (UMA) model of cell survival from a logistic equation. *BJR Open*, **3**(1), 20210040.

[85] Marzi, S, Saracino, B, Petrongari, M, Arcangeli, S, Gomellini, S, Arcangeli, G, Benassi, M, and Landoni, V. 2009. Modeling of alpha/beta for late rectal toxicity from a randomized phase II study: Conventional versus hypofractionated scheme for localized prostate cancer. *Journal of Experimental & Clinical Cancer Research*, **28**(1), 117–124.

[86] McMahon, S J. 2019. The linear quadratic model: Usage, interpretation and challenges. *Physics in Medicine and Biology*, **64**(1), 01TR01.

[87] Mišić, V V, and Chan, T C Y. 2015. The perils of adapting dose to errors in radiation therapy. *PLoS One*, **10**(5), e0125335.

[88] Mizuta, M, Takao, S, Date, H, Kishimoto, N, Sutherland, K L, Onimaru, R, and Shirato, H. 2012. A mathematical study to select fractionation regimen based on physical dose distribution and the linear-quadratic model. *International Journal of Radiation Oncology*Biology*Physics*, **84**(3), 829–833.

[89] Mundt, A J, and Roeske, J C. 2005. *Intensity modulated radiation therapy: A clinical perspective*. Hamilton: BC Decker.

[90] Neumark, S. 1965. *Solution of cubic and quartic equations*. Oxford: Pergamon Press.

[91] Nourollahi, S. 2018. Convex and robust optimization methods for modality selection in external beam radiotherapy. PhD thesis, University of Washington, Seattle.

[92] Nourollahi, S, Ghate, A, and Kim, M. 2018. Robust modality selection in radiotherapy. Pages 11–20 of: Yang H, and Qiu, R (eds) *Proceedings of the INFORMS International Conference on Service Science*.

[93] Nourollahi, S, Ghate, A, and Kim, M. 2019. Optimal modality selection in external beam radiotherapy. *Mathematical Medicine and Biology*, **36**(3), 361–380.

[94] O'Rourke, S F C, McAneney, H, and Hillen, T. 2009. Linear quadratic and tumour control probability modelling in external beam radiotherapy. *Journal of Mathematical Biology*, **58**, 799–817.

[95] Qi, X S, Yang, Q, Lee, S P, Li, X A, and Wang, D. 2012. An estimation of radiobiological parameters for head-and-neck cancer cells and the clinical implications. *Cancers*, **4**(2), 566–580.

[96] Søvik, A, Malinen, E, Skogmo, H K, Bentzen, S M, Bruland, Ø S, and Olsen, D R. 2007. Radiotherapy adapted to spatial and temporal variability in tumor hypoxia. *International Journal of Radiation Oncology*Biology*Physics*, **68**(5), 1496–1504.

[97] Rockwell, S. 1998. Experimental radiotherapy: A brief history. *Radiation Research*, **150**(Supplement), S157–S169.

[98] Romeijn, H E, and Dempsey, J F. 2008. Intensity modulated radiation therapy treatment plan optimization. *TOP*, **16**(2), 215–243.

[99] Romeijn, H E, Ahuja, R K, Dempsey, J F, and Kumar, A. 2006. A new linear programming approach to radiation therapy treatment planning problems. *Operations Research*, **54**(2), 201–216.

[100] Royal College of Radiologists. 2019 (March). *Radiotherapy dose fractionation*. www.rcr.ac.uk/system/files/publication/field_publication_files/brfo193_radiotherapy_dose_fractionation_third-edition.pdf.

[101] Saberian, F. 2015. Convex and dynamic optimization with learning for adaptive biologically conformal radiotherapy. PhD thesis, University of Washington, Seattle.

[102] Saberian, F, Ghate, A, and Kim, M. 2015. A two-variable linear program solves the standard linear-quadratic formulation of the fractionation problem in cancer radiotherapy. *Operations Research Letters*, **43**(3), 254–258.

[103] Saberian, F, Ghate, A, and Kim, M. 2016. Optimal fractionation in radiotherapy with multiple normal tissues. *Mathematical Medicine and Biology*, **33**(2), 211–252.

[104] Saberian, F, Ghate, A, and Kim, M. 2016. A theoretical stochastic control framework for adapting radiotherapy to hypoxia. *Physics in Medicine and Biology*, **61**(19), 7136–7161.

[105] Saberian, F, Ghate, A, and Kim, M. 2017. Spatiotemporally optimal fractionation in radiotherapy. *INFORMS Journal on Computing*, **29**(3), 422–437.

[106] Sachs, R K, and Brenner, D J. 1998. The mechanistic basis of the linear-quadratic formalism. *Medical Physics*, **25**(10), 2071–2073.

[107] Saka, B, Rardin, R L, Langer, M P, and Dink, D. 2011. Adaptive intensity modulated radiation therapy planning optimization with changing tumor geometry and fraction size limits. *IIE Transactions on Healthcare Systems Engineering*, **1**(4), 247–263.

[108] Saka, B, Rardin, R L, and Langer, M P. 2014. Biologically guided intensity modulated radiation therapy planning optimization with fraction-size dose constraints. *Journal of the Operational Research Society*, **65**(4), 557–571.

[109] Santiago, A, Barczyk, S, Jelen, U, Engenhart-Cabillic, R, and Wittig, A. 2016. Challenges in radiobiological modeling: Can we decide between LQ and LQ-L

models based on reviewed clinical NSCLC treatment outcome data? *Radiation Oncology*, **11**(67), 1–13.

[110] Shepard, D M, Ferris, M C, Olivera, G H, and Mackie, T R. 1999. Optimizing the delivery of radiation therapy to cancer patients. *SIAM Review*, **41**(4), 721–744.

[111] Sinclair, W K. 1966. The shape of radiation survival curves of mammalian cells cultured in vitro. In: *Biophysical aspects of radiation quality*. Technical Reports Series Number 58. Vienna, Austria: International Atomic Energy Agency.

[112] Sir, M, Epelman, M, and Pollock, S. 2012. Stochastic programming for off-line adaptive radiotherapy. *Annals of Operations Research*, **196**, 767–797.

[113] South, C P, Partridge, M, and Evans, P M. 2008. A theoretical framework for prescribing radiotherapy dose distributions using patient-specific biological information. *Medical Physics*, **35**(10), 4599–4611.

[114] Steele, J M. 2004. *The Cauchy-Schwarz master class: An introduction to the art of mathematical inequalities*. Cambridge: Cambridge University Press.

[115] Strang, G. 2009. *Introduction to linear algebra*. 4th ed. Wellesley, MA: Wellesley-Cambridge Press.

[116] Syed, Y A, Patel-Yadav, A K, Rivers, C, and Singh, A K. 2017. Stereotactic radiotherapy for prostate cancer: A review and future directions. *World Journal of Clinical Oncology*, **8**(5), 389–397.

[117] ten Eikelder, S C M, Ferjančič, P, Ajdari, A, Bortfeld, T, den Hertog, D, and Jeraj, R. 2020. Optimal treatment plan adaptation using mid-treatment imaging biomarkers. *Physics in Medicine and Biology*, **65**(24), 245011.

[118] ten Eikelder, S C M, Ajdari, A, Bortfeld, T, and den Hertog, D. 2021. Conic formulation of fluence map optimization problems. *Physics in Medicine and Biology*, in press.

[119] Thames, Jr., H D, Withers, H R, Peters, L J, and Fletcher, G H. 1982. Changes in early and late radiation responses with altered dose fractionation: Implications for dose-survival relationships. *International Journal of Radiation Oncology*Biology*Physics*, **8**(2), 219–226.

[120] Trotti, A, Fu, K K, Pajak, T F, Jones, C U, Spencer, S A, Phillips, T L, Garden, A S, Ridge, J A, Cooper, J S, and Ang, K K. 2005. Long term outcomes of RTOG 90–03: A comparison of hyperfractionation and two variants of accelerated fractionation to standard fractionation radiotherapy for head and neck squamous cell carcinoma. *International Journal of Radiation Oncology*Biology*Physics*, **63**(Supplement 1), S70–S71.

[121] Tucker, S L. 1984. Tests for the fit of the linear-quadratic model to radiation isoeffect data. *International Journal of Radiation Oncology*Biology*Physics*, **10**(10), 1933–1939.

[122] Unkelbach, J, and Papp, D. 2015. The emergence of nonuniform spatiotemporal fractionation schemes within the standard BED model. *Medical Physics*, **42**(5), 2234–2241.

[123] Unkelbach, J, Craft, D, Salari, E, Ramakrishnan, J, and Bortfeld, T. 2013. The dependence of optimal fractionation schemes on the spatial dose distribution. *Physics in Medicine and Biology*, **58**(1), 159–167.

[124] van Leeuwen, C M, Oei, A L, Creeze, J, Bel, A, Franken, N A P, Stalpers, L J A, and Kok, H P. 2018. The alfa and beta of tumours: A review of parameters of

the linear-quadratic model, derived from clinical radiotherapy studies. *Radiation Oncology*, **13**(1), 96–106.

[125] Verma, V, Shah, C, Rwigema, J-C M, Solberg, T, Zhu, X, and Simone II, C B. 2016. Cost-comparativeness of proton versus photon therapy. *Chinese Clinical Oncology*, **5**(4), 56.

[126] Vogelius, I R, and Bentzen, S M. 2013. Meta-analysis of the alpha/beta ratio for prostate cancer in the presence of an overall time factor: Bad news, good news, or no news? *International Journal of Radiation Oncology*Biology*Physics*, **85**(1), 89–94.

[127] Wang, J Z, Guerrero, M, and Li, X A. 2003. How low is the alpha/beta ratio for prostate cancer? *International Journal of Radiation Oncology*Biology*Physics*, **55**(1), 194–203.

[128] Webb, S. 2019. *Contemporary IMRT: Developing physics and clinical implementation*. Bristol: CRC Press.

[129] Webb, S, and Oldham, M. 1996. A method to study the characteristics of 3D dose distributions created by superposition of many intensity-modulated beams delivered via a slit aperture with multiple absorbing vanes. *Physics in Medicine and Biology*, **41**(10), 2135–2153.

[130] Wheldon, T E, and Amin, A E. 1988. The linear-quadratic model. *British Journal of Radiology*, **61**, 700–702.

[131] Widmark, A, Gunnlaugsson, A, Beckman, L, Thellenberg-Karlsson, C, Hoyer, M, Lagerlund, M, Kindblom, J, Ginman, C, Johansson, B, Björnlinger, K, Seke, M, Agrup, M, Fransson, P, Tavelin, B, Norman, D, Zackrisson, B, Anderson, H, Kjellen, E, Franzen, L, and Nilsson, P. 2019. Ultra-hypofractionated versus conventionally fractionated radiotherapy for prostate cancer: 5-year outcomes of the HYPO-RT-PC randomised, non-inferiority, phase 3 trial. *The Lancet*, **394**(10196), 384–395.

[132] Williams, M V, Denekamp, J, and Fowler, J F. 1985. A review of alpha/beta ratios for experimental tumors: Implications for clinical studies of altered fractionation. *International Journal of Radiation Oncology*Biology*Physics*, **11**(1), 87–96.

[133] Yan, D, Vicini, F, Wong, J, and Martinez, A. 1997. Adaptive radiation therapy. *Physics in Medicine and Biology*, **42**(1), 123–132.

[134] Yang, Y, and Xing, L. 2005. Optimization of radiotherapy dose-time fractionation with consideration of tumor specific biology. *Medical Physics*, **32**(12), 3666–3677.

[135] Yang, Y, and Xing, L. 2005. Towards biologically conformal radiation therapy (BCRT): Selective IMRT dose escalation under the guidance of spatial biology distribution. *Medical Physics*, **32**(6), 1473–1484.

[136] Yu, C, Shepard, D, Earl, M, Cao, D, Luan, S, Wang, C, and Chen, D Z. 2006. New developments in intensity modulated radiation therapy. *Technology in Cancer Research and Treatment*, **5**(5), 451–464.

[137] Zaider, M. 1998. There is no mechanistic basis for the use of the linear-quadratic expression in cellular survival analysis. *Medical Physics*, **25**(5), 791–792.

[138] Zaorsky, N G, J D Palmer, Hurwitz, M D, Keith, S W, Dicker, A P, and Den, R B. 2015. What is the ideal radiotherapy dose to treat prostate cancer?:

A meta-analysis of biologically equivalent dose escalation. *Radiotherapy and Oncology*, **115**(3), 295–300.

[139] Zietman, A L, Desilvio, M L, Slater, J D, Rossi, C J, Miller, D W, Adams, J A, and Shipley, W U. 2005. Comparison of conventional-dose vs high-dose conformal radiation therapy in clinically localized adenocarcinoma of the prostate: A randomized controlled trial. *Journal of American Medical Association*, **294**(10), 1233–1239.

[140] Zilli, T, Scorsetti, M, Zwahlen, D, Franzese, C, Förster, R, Giaj-Levra, N, Koutsouvelis, N, Bertaut, A, Zimmermann, M, D'Agostino, R G, Alongi, F, Guckenberger, M, and Miralbell, R. 2018. ONE SHOT – Single shot radiotherapy for localized prostate cancer: Study protocol of a single arm, multicenter phase I/II trial. *Radiation Oncology*, **13**(1), 166.

Index

Printed in the United States
by Baker & Taylor Publisher Services